Gendered Approaches to Spatial Development in Europe

This book explores the extent to which gendered approaches are evident and effective in spatial development in selected European countries.

Beginning with an introduction to theories and concepts of gender, space and development, the book includes a brief historical review of gender in spatial planning and development throughout Europe in general, and an overview of different national frameworks in European countries, comparing legal, organisational and cultural similarities and differences. This is followed by a critical reflection on how simplifications and stereotypes of gender concepts are used in the practice of spatial development. The main part of the book offers a transnational discussion of planning practices on selected thematic topics. It starts with gender-sensitivity in urban master planning and at neighbourhood level referring to different types of planning manuals. Furthermore, the book focuses on gender-sensitive evaluation in urban planning as well as international agendas for sustainable development as a framework for a new generation of gender equality policies. The chapter authors assert that climate change, migration and austerity have threatened gender equality and therefore spatial development needs to be especially alert to gender dimensions. The editors end with an outlook and suggestions for further action and research on gender issues in spatial development.

With inputs from some of Europe's leading thinkers on gender, space and development, this volume is designed to inspire students, scholars and practitioners to reflect upon the contribution that gendered approaches can make in the various fields of spatial development and environmental planning.

Barbara Zibell is Professor of Sociology in Architecture and Women's Studies, Head of the Department of Sociology of Planning and Architecture and President of the Forum for Gender Competence in Architecture Landscape Planning (gender_archland) at the Leibniz Universität Hannover, Germany.

Doris Damyanovic is Associate Professor and Deputy Head at the Institute of Landscape Planning, Department of Spatial, Landscape and Spatial Science at the University of Natural Resources and Life Sciences in Vienna, Austria.

Ulrike Sturm is Professor and Head of the Institute of Sociocultural Development at the School of Social Work and Co-Head of the Interdisciplinary Cluster Spatial Development & Social Cohesion at the Lucerne University of Applied Sciences and Arts, Switzerland.

Routledge Studies in Gender and Environments

With the European Union, United Nations, UN Framework Convention on Climate Change, and national governments and businesses at least ostensibly paying more attention to gender, including as it relates to environments, there is more need than ever for existing and future scholars, policy-makers, and environmental professionals to understand and be able to apply these concepts to work towards greater gender equality in and for a sustainable world.

Comprising edited collections, monographs and textbooks, this new *Routledge Studies in Gender and Environments* series will incorporate sophisticated critiques and theorisations, including engaging with the full range of masculinities and femininities, intersectionality, and LBGTIQ perspectives. The concept of 'environment' will also be drawn broadly to recognise how built, social and natural environments intersect with and influence each other. Contributions will also be sought from global regions and contexts which are not yet well represented in gender and environments literature, in particular Russia, the Middle East, and China, as well as other East Asian countries such as Japan and Korea.

Series Editor: Professor Susan Buckingham, an independent researcher, consultant and writer on gender and environment related issues.

Titles in this series include:

Ecological Masculinities
Theoretical Foundations and Practical Guidance
Martin Hultman and Paul M. Pulé

Gendered Approaches to Spatial Development in Europe
Perspectives, Similarities, Differences
Edited by Barbara Zibell, Doris Damyanovic and Ulrike Sturm

Gendered Approaches to Spatial Development in Europe

Perspectives, Similarities, Differences

Edited by Barbara Zibell, Doris Damyanovic and Ulrike Sturm

Routledge
Taylor & Francis Group
LONDON AND NEW YORK

earthscan
from Routledge

First published 2019 by Routledge

2 Park Square, Milton Park, Abingdon, Oxon, OX14 4RN
605 Third Avenue, New York, NY 10017

Routledge is an imprint of the Taylor & Francis Group, an informa business

First issued in paperback 2020

British Library Cataloguing-in-Publication Data
A catalogue record for this book is available from the British Library

Library of Congress Cataloging-in-Publication Data
A catalog record has been requested for this book

ISBN: 978-1-138-58766-3 (hbk)
ISBN: 978-0-367-73079-6 (pbk)

Typeset in Goudy
by Wearset Ltd, Boldon, Tyne and Wear

Supported by

AKADEMIE FÜR RAUMFORSCHUNG UND LANDESPLANUNG

Leibniz Universität Hannover

Universität für Bodenkultur Wien
University of Natural Resources and Life Sciences, Vienna

Lucerne University of Applied Sciences and Arts
HOCHSCHULE LUZERN
FH Zentralschweiz

Contents

Figures

Tables

Boxes

Contributors

Susan Buckingham is an independent researcher, writer and consultant, following 30 years as a university academic, most recently as a professor in the Centre for Human Geography at Brunel University, London. She is interested in the role that gender plays in climate change (and vice versa), and in the relationships between gender and environment. She has volunteered with women's organisations, including as a trustee and chair of the UK's "Women's Environmental Network", and integrates these experiences into her writing. She has also worked with UK-based planning organisations (such as the Royal Town Planning Institute, the Local Government Association) international organisations (UNEP/UNDP) and on overseas projects in Europe (Horizon 2020 projects and with the German government-funded Academy for Spatial Research and Planning ARL), the USA and Pakistan. She is currently working on a second edition of *Gender and Environment*, first published in 2000, and is series editor for Routledge's "Gender and Environments" book series.

Doris Damyanovic is Associate Professor of Sustainable Landscape Planning and Gender Planning at the University of Natural Resources and Life Sciences in Vienna (BOKU Vienna). She studied Landscape Planning and focused in her PhD thesis on "Implementation of gender mainstreaming in spatial planning". In 2016 she finished her professorial thesis (habilitation) with the title "A gender-sensitive approach in landscape planning". For the last 15 years she has conducted research on gender and diversity planning topics in urban and rural areas. She is a member of gender_archland (Forum of Gender Competence in Architecture Landscape Planning) and we4DRR network (European network of women exchange for Disaster Risk Reduction). Her research and research-led teaching focus on urban planning and spatial development, in particular on planning theories, tools and methods and participatory planning. Further research topics are climate change adaption and mitigation strategies in the context of green infrastructure planning and disaster risk reduction.

Sylvette Denèfle is Professor Emerita of Sociology at the François Rabelais University, Tours, and Research Supervisor in UMR7324 CNRS "Cités,

Territoires, Environnement et Sociétés", the scientific lead in several European and national programmes and a scientific expert on various research programmes. The main focus of her research has been the analysis of trends in social norms, and more particularly the analysis of changes linked to the advent of modernity. Her main work consists of research projects looking at the social dimensions of architecture, urban planning and the environment, and specifically the gender perspectives of these topics. Recently, her work has involved setting up research projects looking at what women specifically bring to urban development through professional integration, particularly in a sustainable perspective.

Liisa Horelli, PhD in Environmental Psychology, is Adjunct Professor at the Department of Built Environment, Aalto University. She has also been Visiting Professor at the Politecnico di Milano for designing smart cities. She is a popular keynote speaker at gender conferences. She has carried out action research and published several articles on participatory urban and regional planning, especially from the perspective of health promotion and the enhancement of human- and gender-sensitive environments. Currently, she is studying self-organisation around smart cities and its gendered consequences for urban planning, community development and local co-governance. Dr Horelli has also conducted evaluations of EU-funded programmes and projects from a gender perspective since the 1990s. She is a former president of the Finnish Evaluation Society (FES) and a board member of the European Evaluation Society (EES).

Sandra Huning is a lecturer and researcher at the Faculty of Spatial Planning, Dortmund University of Technology. She studied Spatial Planning in Dortmund, Germany, and Grenoble, France, and completed her PhD at Berlin University of Technology. Her research focuses on the link between urban gender studies, planning sociology and planning theory. She was a member of the International Working Group "Gender in Spatial Development" at the German Academy for Spatial Research and Planning (ARL). Currently she is involved in research on the challenges for socio-environmental policies in the context of housing and on new ways to involve migrant communities in urban development. Dr Huning is co-editor of the planning theory book series "Planungsrundschau".

Eva Kail studied Urban and Regional Planning at the Technical University of Vienna. She has worked for the City of Vienna since 1986. She was the first Head of the Women's Office (1991–1998) and built up and ran the Co-ordination Office for Planning and Construction Geared to the Requirements of Daily Life and the Specific Needs of Women (1998–2009). Since 2010 she has worked as the Gender Expert in the Executive Group for Construction and Technology – Competence Centre overall urban planning, Smart City Strategy, participation, gender planning. She has coordinated about 60 pilot projects in the fields of housing, mobility, public space, urban development and social infrastructure and is one of the leading administrative experts for gender planning in Europe.

Anastasia-Sasa Lada is an architect, Professor Emerita in Architectural Design and Gender Studies, School of Architecture, Aristotle University of Thessaloniki (AUTH). Her main teaching and research interests, since 1985, have focused on the topics of gender, sexuality, architecture and urban space. She has published a number of books and many articles in scientific magazines. She was a founding member of the first Women's Studies Group in Greece, AUTH (1984–1992), an elected Member on the AOIFE Council (2000–2006) and President of the Committee of Gender Equality in AUTH (2008–2013). She was scientific supervisor and project manager of the research project "The concept of gender in social and natural sciences, contemporary research and teaching issues in Greece" funded by the Ministry of Education, Greece (2004–2008), Scientific Consultant in the research project "Mapping the Scientific Space of Greek Women Research Staff", National Documentary Centre, and General Secretary of Research and Technology, Greece (2006–2008).

Tanja Mölders studied Environmental Sciences at Leuphana University in Lüneburg, Germany. After her studies, she worked in various research projects on gender, space and the environment at Leuphana University and the University of Hamburg. She finished her PhD in Social Sciences in 2009 and became the head of the research group "PoNa – Shaping Nature: Policy, Politics and Polity" at Leuphana University (2008–2014). Since 2013 she has held the chair "Space and Gender" at the Faculty of Architecture and Landscape Sciences at the Leibniz Universität Hannover, where she co-operates with the Forum for Gender Competence in Architecture Landscape Planning (gender_archland). Currently, she holds a deputy professorship for Environmental Planning at Leuphana University. In 2017 she submitted her professorial thesis (habilitation), where she focused on the nexus of gender, nature and rurality. She is a member of the German Gender Studies Association and the Academy for Spatial Research and Planning (ARL).

Inés Novella Abril holds a Master's degree in Architecture and in Equal Opportunities. She is a researcher at the Department of Urban Planning and a member of the UNESCO Chair on Gender at Universidad Politécnica de Madrid (UPM). Her research focuses on gender-sensitive architecture and urban planning, and on structural change and gender policies in STEM (Science, Technology, Engineering & Math) organisations. She has participated in various European projects on gender equality and responsible research and innovation (RRI) in research organisations. She has been a guest researcher and guest lecturer at various European universities. She has been in charge of a number of training courses on gender equality in STEM fields targeting professionals, researchers and managers from both public and private institutions. She is the Western Europe Area Editor of the Bloomsbury

Global Encyclopedia of Women in Architecture, to be published in 2021. Her participatory action research with women on safety in public space has been recently selected for the Spanish Pavilion in the 2018 edition of the Venice Biennale of Architecture.

Florian Reinwald is a senior scientist at the Institute of Landscape Planning, Department of Landscape, Spatial and Infrastructure Sciences, at the University of Natural Resources and Life Sciences in Vienna (BOKU Vienna). He studied Landscape Planning at BOKU and TU Munich/Weihenstephan. He completed his PhD thesis at BOKU Vienna in 2017. The focus of his research lies on settlement and community development, planning instruments and planning processes, urban green infrastructure and climate change, as well as village renewal and participation processes including new technologies. He has a special research interest in gender and justice regarding participation, spatial and urban planning and development, also in the context of climate change adaption and mitigation strategies.

Marion Roberts is Emeritus Professor of Urban Design at the University of Westminster, London, UK. She studied Architecture at University College London and worked as an architect before researching for her doctorate. Her PhD thesis was on gender divisions and housing design, later published as *Living in a Man-Made World* (Routledge, 1991). She joined the University of Westminster in 1989 and was promoted to Professor in 2005. She has led research projects on the night-time economy, funded by the Leverhulme Foundation, the Joseph Rowntree Foundation and central and local government. She has maintained her interest in gender issues; she co-led a working group on Cities for the European Commission funded COST research network, genderSTE (2012–2016), co-edited *Fair Shared Cities: The Impact of Gender Planning in Europe* (Ashgate, 2013). She has given talks on engendering cities at conferences and universities in the UK and Eastern Europe and was a Distinguished Visitor to the University of New South Wales, Australia.

Inés Sánchez de Madariaga is UNESCO Chair on Gender and Professor of Urban Planning at Universidad Politécnica de Madrid. She is a member of the Leadership Council of the Spanish UN-Sustainable Solutions Development Network and Advisor to the Executive Director of UN-Habitat. She was a member of the European Commission Helsinki Group on Gender in Research and Chair of the EC Expert Group on Structural Change of Research Organisations. She has been Chair of the COST network genderSTE and Co-Director of the EU-US Gendered Innovations Project. She has directed and advised over 40 research projects, including the European projects Genport, Genovate, RRITools, Libra, Sagerip, GEDII, Gender-Net, RRING and TRIGGER. She is an EC Urbact expert and a member of the Scientific Committee of the UNESCO project SAGA. She has held public office in Spain as Deputy Director General for Architecture, Advisor to the

Minister of Housing, Advisor to the Minister of Science, Director of the Women and Science Unit at the Secretary of State for Research and Development. She has been a Visiting Scholar at Columbia University in the City of NY, the London School of Economics and the Bauhaus School of Architecture.

Ulrike Sturm has been a lecturer and researcher at the Lucerne University of Applied Sciences and Arts (LUASA) since 2010. From 2013 to 2018 she was head of the Interdisciplinary Focus Area "Cooperation in Building and Planning". In 2018 she became professor and head of the Institute of Sociocultural Development at the Department of Social Work and launched, with architect Stefan Kunz, the Interdisciplinary Cluster "Spatial Development & Social Cohesion" at LUASA. After a Master's degree in Philosophy and Comparative Literature at the Free University of Berlin in 1992, Dr Sturm received her diploma as an architect at the Technical University of Berlin in 1998. She received her PhD in the Theory and History of Urbanism in 2011. For over ten years she gained practical experience as a freelance architect working for various offices of architecture and urbanism as well as in her own practice. From 2000 on Dr Sturm worked as a teaching assistant in urbanism at the Brandenburg Technical University Cottbus. From 2008 to 2010 she was Assistant Professor at Université de Quebec à Montreal and Leibniz Universität Hannover.

Stephanie Tuggener is project leader at Kontextplan AG, a company for transport planning and community development. She studied Geography at the University of Zurich (major) and urban planning at ETH Zurich (minor), Switzerland. During her studies she undertook a traineeship at Aalto University, Helsinki, and worked as a research assistant in various projects on space and politics at SOTOMO (research centre for society, politics and space). She continued as an urban planner at the Department of Spatial Planning, Canton of Aargau (2012–2016), Switzerland. Since the summer of 2017 she has worked at Kontextplan AG in Zurich, Switzerland. Stephanie Tuggener is a board member of Lares – association for gender-sensitive planning. Between 2016 and 2018 she took part in the Academy for Spatial Research and Planning's (ARL) mentoring programme.

Lidewij Tummers, Ir. (Dipl.-Ing.) wrote her PhD thesis on co-housing, spatial planning and energy transition, entitled "From passive house to active inhabitants: learning from co-housing pioneers". The study applies a combined framework of gender and niche-innovation theory, analysing the engineering of collaborative housing projects. A founding member of the international network of co-housing researchers, she continues to research commoning the engineering of low-impact housing. Since 2006 she has been a part-time tutor-researcher at the Department of Urbanism, Faculty of Architecture and the Built Environment, TU Delft, visiting lecturer at several European universities and "Gender, Science and Technology" trainer for early stage researchers, post-docs and professionals. Her work and publications focus on sustainable housing, specifically self-managed housing and energy transition as well as gendered perspectives in spatial development and criteria for inclusive urban design. Based at the Rotterdam studio Tussen Ruimte (Intermediate Space), independent designers and technical consulting engineers, she advises municipal and regional authorities on integrating residents' initiatives and gender concepts in sustainable spatial development, and residents' associations on sustainable, energy-efficient building and refurbishment.

Heidrun Wankiewicz is a planner, gender expert and part-time researcher, and since 2003 director of her studio planwind.at – planning.management.research (Salzburg, AT). After completing her Master's thesis in Geography and French on urban planning (City of Strasbourg) and on regional governance (France meridional), she specialised in participatory local and regional development and local land-use planning, with a focus on inclusive spatial strategies. Since 2016 she has held a PhD in Social Geography based on her applied research on gender mainstreaming in spatial planning and regional development. Her research interests include transdisciplinary settings for gender-fair planning processes and the implications of demographic change for a shift in governance towards the co-creation of infrastructures of everyday life at regional scale. She works in regional, national and international consortia and in regional and European networks, such as gender_archland, Frauenrat Salzburg and ARL-International Working Group. In 2018, with Lidewij Tummers, she launched the European agency corrina.eu – Co-housing, Regional Research & Innovation Agency.

Barbara Zibell studied Urban and Regional Planning at the Technical University of Berlin, Germany, and became a government-licensed urbanist in 1984. In between, from 1984 to 1988, and during her academic employment at the Institute of Technology in Zurich, 1989–1996, where she completed her PhD, she worked as a consultant and evaluator as well as a practitioner in spatial planning and development, both in Switzerland and Germany. In 1996 she became Professor of Sociology in Architecture and Women's Studies first, later Head of the Department of Sociology of Planning and Architecture at the Leibniz Universität Hannover, where she has been teaching and doing research since then. Barbara Zibell is initiator and head of the international working group "Gender in Spatial Development", which was set up by the German Academy for Spatial Research and Planning (ARL), of which she has been a member since 2001. She is also initiator and president of the Forum for Gender Competence in Architecture Landscape Planning (gender_archland) at the Faculty of Architecture and Landscape Sciences at the Leibniz Universität Hannover.

Forewords

> The Academy for Spatial Research and Planning (ARL) – Leibniz Forum for Spatial Sciences, aspires to be a forum and centre of competence in sustainable spatial development. Its work is characterised by holistic, integrative and future-oriented perspectives on complex spatial-related societal challenges.
>
> (Translated from the research concept of the ARL 2017–2022)

Establishing a Working Group within the Academy for Spatial Research and Planning (ARL) frequently requires a lengthy preparation period and calls for intensive – sometimes even controversial – discussions. This was also the case prior to the establishment of the International Working Group (IWG) on "Gender in Spatial Development – Perspectives, Similarities and Differences", which gave rise to this volume. In my former role as Vice-President of the ARL, in 2013 and 2014, I was intensively involved in these discussions. As resources are scarce, important questions face decision-makers. The following were perhaps particularly significant: Is the proposed subject an important and urgent problem of major relevance to spatial planning and society? Does collaboration between (in this case international) academics and practitioners in the working group promise to generate answers to one of the great challenges of the present day? And what new findings – going beyond the ARL's research to date – for spatial research and planning practice can be expected from the collaboration of experts in the context of the working group?

There were no easy answers to these three questions when considering the subject proposed by Barbara Zibell and colleagues – the investigation of the contribution of gender as a category to a sound, critical understanding of processes of spatial development and to achieving sustainability objectives for future spatial development. Does the use of a gender framework in the spatial and planning sciences really represent a new and topical field of research – even though "Gender Planning" is a concept that has been widely discussed among academics for more than 40 years and "Gender Mainstreaming" has been a binding European strategy for every form of public governance and consequently also for spatial design for almost 20 years? Doubts about the topicality and urgency of the proposed subject therefore, initially, seemed to be appropriate.

These findings by the IWG now sweep aside every doubt about the relevance, urgency and usefulness of the research question that the 15 colleagues in the sphere of spatial and planning sciences and practice have pondered against the background of various European planning cultures. Perspectives from human geography and environmental sciences, architecture and urban design, landscape and spatial planning, urban and regional planning, sociology and environmental psychology have been incorporated through interdisciplinary teamwork. After (almost) four years of intensive and constructive collaboration, the members of the IWG have now presented a publication that answers the above questions soundly and clearly. Yes, heuristic questioning of spatial development through the gender category is unequivocally topical, precisely in the face of the situations of social inequality that have developed rapidly to crisis point in recent times – and is extremely wide-ranging. "Gender" acts as an eye opener to various socio-ecological inequalities and relationships of power and authority structures, and how they can be mapped within spaces. This means that the aspiration to control spatial planning and spatial development in accordance with the criterion of fairness can be grasped and shaped in a way that is viable for the future.

From a gender perspective – as is also demonstrated by the retrospective overview of the conceptual development of "Gender Planning" presented in this volume – it becomes clear just how great a challenge is bound up specifically with sustainable spatial development. Planning modes are required that focus on intra- and inter-generational fairness and can combine social, ecological and, from a real-world everyday life perspective, economic development objectives and integrated solutions to problems. Sustainable spatial development needs the kind of constructively critical and future-oriented perspectives that are brought together in this volume.

I hope that this book enjoys a broad readership among (spatial) science academics and practitioners, and receives critical international acclaim.

Sabine Hofmeister
Former Vice-President of ARL, former Professor for Environmental Planning at Leuphana University of Lüneburg

Feminists have long been critical of conventional planning practice for excluding gender concerns from serious consideration. Rooted in universalist Enlightenment thinking, early planners had regarded society as a unitary whole with a single public interest. They considered themselves to be technical experts whose prescriptions would benefit all, considered themselves to be non-political, and did not view top-down planning as contravening democratic values. Men dominated the physical planning profession, and they assumed that separation of home and work was appropriate. The home, under the oversight of the wife/mother, constituted a haven from the competitive world of factory and office. This conception became enshrined in zoning that segregated uses, reflecting widely accepted binaries between production and reproduction, public and private. The postwar introduction of cost-benefit analysis into planning processes further intensified the bias towards patriarchy, since attaching a monetary value to the impacts of spatial planning meant that returns on public investment were calculated based on the superior wages earned by men. In other words, transporting men to work produced greater aggregate gains than moving women to part-time jobs, schools or shopping. Only with the rise of second-wave feminism in the 1960s and 1970s did planning scholars challenge the implicit male bias within the concept of the general good. It was even later when more fluid gender conceptions that did not divide the world into male and female took hold.

Critique alone, however, does not produce guidance for planners seeking greater inclusion of women and LBGTQ (Lesbian, Bisexual, Gay, Transgender, Queer) individuals in developing their priorities. The editors of this volume seek to overcome this gap, addressing how a "gender+" perspective can become integrated into the structures that govern spatial planning. They begin with a social construction understanding of the ways in which masculine, feminine and gay behaviours are defined, thereby defining gender roles as malleable. After exploring the theoretical context from which a gendered view of space derives, contributors to the book examine the ways in which a gendered perspective translates into planning doctrine within various European countries and metropoles. While calling for specific policies that acknowledge difference, the authors avoid stereotyping female activities. They thus move away from seeing an eternal sexual division of labour based on men as providers and women as caretakers with only a secondary labour market function. Their ultimate purpose is to mainstream gender rather than to compartmentalise it as a special type of spatial planning. This means a move away from designing spaces for the traditional nuclear family and towards producing opportunities for collective activities that reduce the burden of housekeeping and overcome the division between home and work.

The empirical chapters within the book investigate how practice has at times incorporated sensitivity to the aim of gender equality. Using this approach, the authors examine how the various centralised (i.e. top-down) and decentralised planning structures within Europe translate into programmes that enhance gender equity. In particular, they cite Vienna as providing models of gender-sensitive programmes. Although they do not regard the task of mainstreaming gender as nearly accomplished, they do identify examples of progress in that direction even while also seeing the ways in which power is used to protect the interests of the already empowered. They stress both process and outcome; they analyse both neighbourhood and nation as policy-setting authorities, but also argue that all are subject to global phenomena including climate change and neoliberal ideology. Thus, this is a book that, while focused on European spatial planning, allows broader generalisation to the need everywhere for sensitivity to the present vulnerabilities of marginalised groups and to the potential for their transformation into equal participants in framing plans and equal beneficiaries of planning outcomes.

Susan S. Fainstein
Senior Research Fellow, Harvard University Graduate School of Design

Preface

Why gender still matters in spatial development

Barbara Zibell, Doris Damyanovic and Ulrike Sturm

This book is the result of collaboration between European academics and planning practitioners, combining interest in gender perspectives with the theory and practice of spatial development. It has emerged from a process that began in 2014, concluding – for the time being – in the presentation of this volume.

The question as to *why gender still matters* (with reference to spatial development) is of general interest – as is, in particular, the motivation for this collaboration between experts in spatial development and planning. In the light of the (relatively) prolonged efforts that followed the introduction of gender mainstreaming as an international strategy, first at the Third Women's Conference in Nairobi in 1985 and then in a European context in 1998 by the Council of Europe, the common perception of these planning experts is sobering. In particular, the degree of integration and implementation in spatial theory and practice of the existing body of knowledge deriving from gender studies and gender planning is not as widespread as had been hoped.

The "Gender in Spatial Development" research framework is based on a threefold concept reflecting different areas, levels and degrees of:

- Knowledge transfer: why and how are gendered projects/processes started at all?
- Knowledge integration: how, when and under what conditions can gendered projects be anchored in given structures?
- Knowledge implementation: what kinds of structures allow gender-oriented projects and processes to become a permanent integral part of spatial development?

This triangular concept can be used to analyse and discuss the conditions of gendered sustainability in spatial development with regard to planning processes, outcomes and institutional structures in relation to required actions and measures. A comprehensive approach, i.e. one that takes into account all three aspects, is a precondition for the successful and sustainable implementation of gender in spatial development – as confirmed, for instance, by an international study in the context of the Interreg project GenderAlp! by Zibell, Dahms and Karácsony, published by the Province of Salzburg in 2006.[1]

The international working group – formation and members

The institutional framework for collaboration was provided by an international working group (IWG) of the German Academy for Spatial Research and Planning (ARL) on the subject of "Gender in Spatial Development – Perspectives, Similarities and Differences". International working groups were introduced relatively recently at the ARL, and bring together academics and practitioners from various countries to work over a limited period of three to four years on current specialist issues of common interest.

The impetus for an IWG on this particular issue was provided by the International Conference "Theorising and Practising Gender-Sensitive Planning in European Discourse", which took place in April 2012 at the Forum for Gender Competence in Architecture and Landscape Planning (gender_archland) at the Leibniz Universität Hannover. Certain authors of this book as well as representatives of the ARL (Sabine Hofmeister, Evelyn Gustedt) took part. Some of the international participants were recruited from the European network "Gender and Diversity in Urban Sustainability" (GDUS), a platform for academic exchange on common areas of research and practical experience created in 2007, from which, for instance, the publication *Fair Shared Cities*[2] emerged. Some of the participants and members of the subsequent IWG had also collaborated previously in the aforementioned Interreg project "GenderAlp! Spatial development for women and men" (2004–2007). A basis for mutual understanding and a joint body of knowledge and experience thus already existed, nevertheless, collaboration in the IWG – particularly in the early meetings – repeatedly brought new challenges.

The establishment of an international working group with an emphasis on gender was a novelty for the ARL and so also presented a particular challenge for the initiators: never before had one of ARL's international or even national working groups espoused this particular critical perspective. The results presented here, and their publication by one of the most notable European publishing houses in the sphere of planning discourse and theory, is therefore of particular importance to the ARL, which, as a leading academic institution in Germany, has a long and successful tradition of promoting exchange between spatial planning theory and planning practice and has, in recent years, intensified its focus on international collaboration.

The members of the working group include researchers and planning practitioners, recruited through a call for membership. They represent nine European countries: Germany, Austria, Switzerland, Finland, France, Great Britain, Greece, Spain and the Netherlands. As well as 11 universities, the City of Vienna with its longstanding experience in gender-oriented city planning and development is also included, as are planning offices from Austria and the Netherlands with international and transnational research competences in gender-related issues of spatial development. The members represent not only a cross-section of various European countries with their different political systems and planning cultures, but also a considerable spectrum of disciplines, including human geography and environmental sciences, architecture and urban design, landscape and spatial planning, urban and regional planning, sociology and environmental psychology.

Although the IWG does not cover the entire European area – the lack of response to the call for membership shown by colleagues from Eastern European countries is to be regretted – it does represent a cross-section of the old Western European countries. In terms of professional discipline, national background, cultural origins and mindsets, the IWG members are thus characterised by a high degree of diversity. Common to all members is an academic and/or planning practice interest in spatial developments (urban and rural) from the point of view of gender studies, considered to comprise critical perspectives which bring an innovative contribution to the advancement of the spatial planning disciplines involved. The common feminist identification is also shared by the only male member of this 12-strong group.

The motivation to cooperate, on the one hand, was driven by a desire to see the notion of gender perspectives respected and thoroughly integrated in mainstream planning theory and planning and development practices. On the other hand, there was also a wish to pass the existing body of knowledge on to the next generation. Key questions were:

- How do we as practitioners and academics go about equipping the next generation of professionals with relevant skills and tools?
- What worked and what didn't – why?

These shared aims defined the target audience for this volume: planning experts and spatial development academics, students of various disciplines concerned with spatial planning, and those in relevant positions of political and administrative responsibility.

Starting point, process and debates

The academic framework was supported by the concept of sustainable spatial development with its two normative elements: the concept of justice (intra- as well as intergenerational) and the concept of integration. It was regarded as incontrovertible that justice between women and men, especially equality of paid and unpaid labour, shaped by symmetrical gender relations, is a prerequisite for sustainable development. This idea is backed by the Treaty of Amsterdam, which, since 1997, has required that the "Community shall aim to eliminate inequalities, and … promote equality, between men and women" (Art. 3 Para. 2), and by the strategy of gender mainstreaming based on this article, which has since been introduced in all member states. Nonetheless it has to be stated that – not only in the professional world of spatial development and planning practice and research – the contents of the Treaty have not yet become an accepted standard and are far from being thoroughly integrated in planning practices and academic reflection.

These key ideas and observations led to the guiding question of the IWG debates: "How does recognising gender relevance impact on the organisation and structure of land-use planning and appropriation in different European cultures?" Pursuing this question, the IWG first worked on a theory-based research framework. This became the starting point for designing a gender-oriented survey focusing on specific patterns in (control) mechanisms of land use, planning and development in the European countries represented in the group. Based on this survey, in a comparative analysis, differences and deficiencies were identified and described. This subsequently led to new research questions.

The contents and the chapters of this volume have been generated over several years in a process involving seven meetings held throughout Europe. Open discussion, in some cases inspired and enriched by local visits, as well as collaborative learning processes and work on concrete subjects in mixed groups, characterised the mode of cooperation. The work on the chapters of this book was ultimately carried out – depending on the composition of the particular teams of authors – in transnational settings, facilitated by Skype meetings or video conferences. The members of the group have not only broadened scientific findings in the field of research by their participation in the meetings and written contributions, but the transnational exchange of ideas has also helped to provide new insights into national distinctions and international similarities.

Outline of the publication

Crucial points of discussion were the definitions of terms and concepts and their application – concerning gender (mainstreaming) and gender planning, built and social space, and implications of (social) sustainability for gender-sensitive spatial development. An analysis of terms and concepts is incorporated into the common theoretical framework which combines different discourses (Chapter 1) as well as into the brief historical review of spatial planning and development throughout Europe in general (Chapter 2). The authors of the following chapters refer to these terms and concepts and explain their own approaches in the context of their particular topics. This shows that gender concepts and perspectives do not exist in the singular; they are, and will remain, plural and varied. For this reason, several perspectives on gender in spatial development – as well as on spatial development itself – are represented in this book.

After the brief historical introduction, Chapter 2 gives an overview of the different frameworks on a national level as a basis for comparing legal, organisational and cultural similarities and differences in the countries represented within the group. Chapter 3 offers a critical reflection on simplifications and stereotypes in the use of gender concepts in the practice of spatial development, and describes the state of the art in the European countries involved. The thematic Chapters 4–8 discuss five selected topics which have crystallised from the working group debates, from different national perspectives, depending on the provenance of the authors.

One of the most crucial insights is that the intensity with which gender perspectives have a chance to be integrated into mainstream planning depends on the period and the spatial context, level of intervention and project content. Therefore, short-term local and neighbourhood projects may impact on the everyday use of space by adding user-specific qualities focusing on practical needs, but mostly do not change the planning system – with the exception of pilot projects which inspire followers. Regional or district projects which are conceived explicitly as gender projects and refer to gender concepts require strong political will and leadership figures with convincing personalities. The gender perspective in institutions is often lost when the key person leaves. To date it is hard to find true feminist transformation projects which intend – and also achieve – societal change by transforming gender relations. Due to a lack of common or political strategic interests, feminist transformation is still a utopia – or, at any rate, a vision. At the same time, the chance or impact of transformation depends on the particular planning system and economic situation. The following section gives an overview of the content of the chapters in detail.

The chapters in detail

In **Chapter 1** *Sandra Huning, Tanja Mölders* and *Barbara Zibell* provide an overview of concepts and debates around gender, space and development in the European context, acknowledging differences but focusing on commonalities and recent developments. In many contexts, a re-conceptualisation of both space and gender as socially constructed and as process categories can be observed, however, the potential of this constructivist "turn" has not yet been fully exploited. Gender justice in spatial development must move towards an "engendering" of spatial planning and development in more general terms. This is particularly important in pursuit of sustainable spatial development (both as a normative goal and legal obligation), since gender-oriented concepts offer valuable critical perspectives on economic relations, socio-spatial resources, the needs of different target groups, ideals of democracy and unequal power relations in spatial development and planning. This can also offer new insights to expose blind spots in current sustainability debates.

Chapter 2 on international and national perspectives opens with a review by *Doris Damyanovic* and *Barbara Zibell* of the history and future of gendered approaches in planning sciences in nine different European countries, considering different planning scales and levels. In the second section *Stephanie Tuggener* and *Barbara Zibell* then introduce the planning systems of Germany, Austria, Switzerland, Finland, France, Great Britain, Greece, Spain and the Netherlands, identifying gendered approaches. The chapter concludes with insights by *Stephanie Tuggener, Barbara Zibell* and *Ulrike Sturm* gained from international comparison on the one hand, and looks ahead to the thematic chapters on the other hand.

In **Chapter 3** *Lidewij Tummers, Sylvette Denèfle* and *Heidrun Wankiewicz* look at contradictions and challenges in gender mainstreaming and development, suggesting that "gender" is not structurally embedded in urban planning practice, is frequently misunderstood as treating women as a specific target group, and is often "implicit" – lacking explicit standards for (gender) equality. The research identified three types of controversies that arise from gender mainstreaming in planning: the inherent risk of stereotyping fixed identities when defining special needs, tensions between long-term strategies and short-term direct but partial objectives, and the reconciliation of everyday and proximity perspectives with a global perspective. The authors argue that structural embedding of gender mainstreaming produces innovation in planning cultures, challenging planning systems and institutions.

Chapter 4 by *Florian Reinwald, Marion Roberts* and *Eva Kail* examines how gender-sensitive policies and intentions are visible at a strategic city-wide level and in major development sites. The two central research questions are: How has gender mainstreaming been implemented on different spatial development levels such as state level in Austria and England and city level in Vienna and London? How has gender mainstreaming been implemented and which criteria of (implicit) gender-sensitive planning are visible in urban masterplans comparing Barking Riverside (London) and Aspern Seestadt (Vienna)? These two sites are both brownfield urban extensions and in both the aim is to produce a place with coherence and an identity that responds to the demands of everyday life in the twenty-first century. The chapter highlights the critical significance of social infrastructure to a gender-sensitive approach to planning and urban design.

Chapter 5 by *Ulrike Sturm, Stephanie Tuggener, Doris Damyanovic* and *Eva Kail* introduces important planning guidelines at neighbourhood level taking Vienna as an example for Austria, and different guidelines for Switzerland. Various examples of development sites and parks in Vienna and Zurich show to what extent and how the criteria influence planning practices. Finally, the authors summarise the main findings with respect to important gender-sensitive planning principles and criteria in neighbourhood planning and their communication through manuals. It seems that similar criteria can stem from an explicitly gendered approach and from an implicitly gender-sensitive approach focusing on quality of living and urban spaces. However, manuals cannot do "the whole job", strategic gender interests still have to be pursued on different political and planning levels to reach greater gender equality.

Liisa Horelli and *Doris Damyanovic* discuss the challenge of conducting evaluations of urban planning and spatial development from a gender+ perspective in **Chapter 6**. The methodological results of gendered evaluations of two cases in the context of self-organisation around spatial development in Helsinki, Finland, are presented. The authors argue that the complexity of engendering evaluations of urban planning requires an integrated evaluation framework (gender studies, planning literature, feminist/gender evaluation approaches) and the application of mixed methods in order to facilitate knowledge building and to activate "learning to learn" behaviours around urban issues. The results furthermore revealed that evaluations are conditioned by context. Raising awareness and knowledge concerning the quality, equity and justice of the urban and spatial environment are vital.

In **Chapter 7** *Inés Sánchez de Madariaga* and *Inés Novella Abril* discuss how the new international agendas for sustainable development – most significantly the Agenda 2030 and the New Urban Agenda (NUA) – provide a framework within which a new generation of gender equality policies and tools should be advanced in urban planning. The chapter discusses whether existing experiences in integrating gender into urban, land-use and regional planning qualify properly as gender mainstreaming understood as a systematic policy approach encompassing different levels and policy actions. In particular, the chapter looks at the Vienna case, which provides an opportunity to explore key issues, and at the recently adopted Basque Country's Regional Plan, which suggests a possible regional-scale model. In order to fully implement the gender equality goals embedded in the Agenda 2030 and the NUA, more effective gender mainstreaming approaches to spatial planning must be put in place. The chapter concludes by indicating criteria that should be an integral part of efforts to put in place and implement more ambitious agendas for gender mainstreaming in spatial development.

In **Chapter 8** *Susan Buckingham* and *Anastasia-Sasa Lada* identify three problems of global reach that create particular challenges for spatial development and for which countries in the Global North have some responsibility – climate change, migration and austerity. The authors argue that the causes of these problems result from hyper-masculinist structures, which now characterise the whole of the global economy. However, the severest impacts of each are spatially concentrated amongst the poorest and most disadvantaged communities, with particular implications for women. The authors propose that spatial development can change the ways in which it operates to address both the causes and the effects of these problems, and draw on various European examples to illustrate how such changes are being made.

In **Chapter 9** the editors return to the question of how recognising gender relevance impacts on the organisation and structure of land-use planning and appropriation in different European cultures, describing similarities and differences with binary opposites and summarising findings with regard to knowledge transfer, integration and implementation. Some countries deal with gender concerns implicitly within quality-driven planning for everyday needs; however, formal, top-down and explicit approaches with more binding commitment in politics, law and planning are necessary to change power relations. Informal and bottom-up activities can help (re-)introduce needs overlooked in official proceedings. Gender concerns should be on research and policy agendas and part of educational programmes, especially in disciplines relevant for spatial development. The authors are cautiously optimistic that the fourth generation of feminism will increase gender awareness and help to equip professionals with the appropriate skills and tools for gendered spatial development and planning.

Dedication and acknowledgements

With the chapters outlined above, this book gives an insight into the debates of this group of experts and the European countries they represent. Moreover, it can claim to be one of the first publications to present this subject and to raise current debate to an international level. At the same time, it provides a cross-section of European planning systems and their gender impacts and is thus suitable for use as a manual, rendering it of interest to both academia and planning practitioners.

We are grateful to everyone who has helped us to create this volume, in particular:

- the ARL and its representatives – first of all Sabine Hofmeister, at the time of the IWG proposal a member of the ARL executive committee; Dietmar Scholich, at the time Secretary General, followed by Rainer Danielzyk; as well as Evelyn Gustedt and Andreas Klee, members of the executive staff of the ARL Headquarters in Hannover;
- "Forum for Gender Competence in Architecture and Landscape Planning" (gender_archland) at the Leibniz Universität Hannover, for support from its academic staff, in particular Katja Stock, PhD student at the Faculty of Architecture and Landscape Sciences and former member of the Young Professionals' Forum of the ARL, who assisted in the management of the IWG;
- University of Natural Resources and Life Sciences, Vienna (BOKU Vienna), for support given by the Institute for Landscape Planning, Department of Landscape, Spatial and Infrastructure Sciences; and
- Lucerne University of Applied Sciences and Arts for support given by the Interdisciplinary Focus Area "Cooperation in Building and Planning".

We also thank Susan Buckingham, Marion Roberts, Katharine Thomas and Ursula Liebl for their support in editing the book.

Lastly, we want this book to contribute to providing answers to the key question as to *why gender still matters* in spatial development and not to let the debate around this question rest – at least until all genders are accorded equal representation in planning projects and processes and are involved on an equal basis in planning structures, and until gender relations within the structures of society are organised in a spirit of partnership.

Notes

1 Zibell, B. with Dahms, N.-S. and Karácsony, M. (2006). Requirement-oriented Spatial Planning. Gender practice and criteria in spatial planning. Final report. English Abstract. Land Salzburg, Salzburg.

2 Sánchez de Madariaga, I. and Roberts, A. (Eds.) (2013). *Fair shared cities. The impact of gender planning in Europe*. Farnham: Ashgate.

1 Gender, space and development

An introduction to concepts and debates

Sandra Huning, Tanja Mölders and Barbara Zibell

Introduction

Talking about gender, space and development involves exploring inter- and transdisciplinary fields of research and practice which have developed in their particular settings over time and are characterised by different, and at times contradictory, concepts and debates. Specific national trajectories of the women's movements as well as particular planning cultures and systems have had an important impact on the ways in which the concepts and debates around gender, space and development have been framed, adapted and developed in different national contexts (see Chapter 2). Some concepts have travelled across national or regional borders in the Western world and, at first sight, have helped to establish a common understanding among researchers and practitioners. However, the different ways in which such concepts have been employed and embedded in their respective environments make it necessary to re-engage with terminology, professional understandings and socio-cultural, -economic and -political settings to enable cross-European communication with regard to these issues.

This chapter aims to provide an overview of concepts and debates in the European context with a focus on commonalities and recent developments. As three German researchers, we recognise that it is unavoidable that we argue from our own "partial perspective" (see Haraway, 1988), which is shaped by a necessarily limited range of literature, discourses and our particular fields of expertise (spatial science, sustainability science, urbanism and spatial planning). With kind support from our colleagues from North, West and South European countries, we intend to broaden this perspective, while acknowledging that there is a much broader global debate that we will not be able to consider. We do not intend to give definite answers and universal definitions, but instead to show a range of concepts and debates which have proved fruitful for the discussion, can be used in various settings, and are in many cases further considered in later chapters of this volume. Thus, this introduction seeks to fulfil a heuristic function for the following chapters with regard to the categories of gender, space and (spatial) development – including spatial planning – and also with regard to sustainable spatial development as a shared analytic and normative orientation.

Ever since the 1992 Rio declaration, sustainability has been recognised worldwide as a key principle for bringing together environmental and development issues. In order to achieve intra- and intergenerational justice, sustainability means respecting the reproductive requirements of natural resources and attributes a strong role to the participation of women – including the fight against discrimination. Thus, gender perspectives in sustainable development involve both strengthening women within the discourse and pursuing a feminist critique of the discourse (see the last section "Sustainable spatial development – integrating concepts and debates"). Translated into spatial development, the sustainability principle serves not only as a normative guideline that gives orientation to politicians and professionals, but is also regulated by planning legislation and other legal obligations at different territorial levels. Therefore, sustainable spatial development is not only an option but an obligation and responsibility of stakeholders. At the same time, gender equality, equity and justice[1] are much more than optional in development and planning. At both the supranational and national level, especially in the European context, gender concerns and gender justice are to be incorporated into every political programme, plan and measurement. One example is the adoption of the gender mainstreaming principle in 1999 (European Communities, 1997).

This chapter is organised along the three categories that we consider essential for our discussion. Attention is first directed towards the category of gender (the first section). For decades, feminist scholars and practitioners have engaged in a debate that has developed from the assumption of biologically distinct women and men to the notion of gender as a social construction. In order to systematise diverse analytic perspectives, and to match the theoretical and empirical diversity within gender studies and spatial politics, we introduce approaches and differentiations which appear to us to be promising with regard to spatial development. The second relevant category is space (the second section). Spatial sciences traditionally assumed space to be simply given, without much reflection (cf. in Germany the critiques by Sturm, 2000), often understood as surface or territory (geography, spatial planning), or as a container (architecture and urbanism), insofar following an adapted understanding of the ancient world. In the course of the so-called "spatial turn" of the 1980s, space became a central category in social sciences and humanities (notably in sociology and history). Here space is no longer an abstract dimension, but a category exploring relations and (contingent) interpretations. Thus, space is an entity that provides economic, environmental and social living conditions, a notion which underlies this book. The third important category is spatial development as a field of action and governance among politics, planning, economics and people (the third section). Here the category gender becomes relevant with regard to both theory and practice (the fourth section). In this context, the political strategy of gender mainstreaming is also crucial for the concept of gender planning as "gender mainstreaming in spatial planning", but it is certainly not the only path towards gender recognition in planning. Summarising our findings, we focus on sustainable development as a critical concept and transformative project, which allows the nexus of gender, space and (spatial) development to be addressed from a feminist perspective (the fifth section). Thus, we argue that the normative ideal and legal objective of sustainable development must include gender as a key category of societal factors. Sustainable development also connects gender aspects and spatial development both through gender equality and equity requirements (which will be explained in more detail below) and the need to integrate different social, environmental and economic requirements. We claim that in pursuit of sustainable spatial development, gender-oriented concepts challenge today's hegemonic rationalities, question current modes of allocating socio-spatial resources and require the inclusion of marginalised target groups. These critical perspectives are directly linked with visions of alternative economic concepts and ideals of democracy, and their spatial realisation in new forms of housing, the design of public spaces and other spatial plans and development.

Gender

The aim of this book is to look at spatial development from a gender perspective. We argue that the category of gender is of central significance for an understanding and evaluation of spatial structures and processes. Gender perspectives may function as epistemological approaches aiming to make the relevance and scope of gender visible. In recognition of the manifold attempts and proposals to systematise gender approaches in theory and practice (e.g. Harding's 1991 classification of feminist schools of epistemology, Keller's 1995 differentiation for science studies "women in science", "science of gender", "gender in science"), we follow a differentiation of analytic perspectives developed by Sabine Hofmeister, Christine Katz and Tanja Mölders with regard to the gender and sustainability nexus (Hofmeister, Katz and Mölders, 2013; see also Hofmeister and Katz, 2011). To further systematise gender-related concepts and debates we classify the analytic perspectives by distinguishing one biological and two societal gender categories. These perspectives are based on distinctive understandings of the category gender, reflecting the history of women's and gender studies and exploring the theoretical and practical added-value of each particular viewpoint. Each gender category is related to specific methodologies aiming to understand and operationalise the particular perspective. Finally, each gender category tends to correspond to selected policy applications addressing a different gender perspective (see Table 1.1).

Biological category

Gender as a biological category addresses women and men as two different and distinguishable *sexes*. This understanding marks the starting point of a women's studies research approach which was informed by feminist movements and women's policy of the 1970s and 1980s. The declared aim of this kind of analysis is to make the inequality and discrimination of women, as a clearly distinguishable

Table 1.1 Analytic categories of gender with related methodologies and applications

Category	*Meaning*	*Related methodologies*	*Policy and planning applications*
Biological category (Sex)	Biological/ physiological characteristics defining female or male	Quantitative methods + enabling tools to empower participants	Promotion of women; women's equality; women-friendly, women-adequate planning
Societal categories			
• Structural category (Gender)	Structural hierarchies and gender contracts impacting activities in space and time	Qualitative theories, concepts, patterns and methods of analysis; enabling methods to encourage participation; quantitative methods for complexity management	Gender mainstreaming; gender (-sensitive/ -aware) planning
• Process category (Queer, gender identity, expression, orientation)	(Un)Doing and deconstructing gender+ in context as a psycho-social and socio-cultural trajectory leading to varying gender expressions and sexual orientations	Psycho-social theories; queer theories; deconstruction; transcending binary and fixed categories	Gender and diversity management; inclusive/ communicative/ performative planning; deconstruction in planning

Source: based on Hofmeister, Katz and Mölders, 2013; Horelli and Wallin, 2013; Roberts, 2013.

collective subject, visible and criticisable. A great advantage of this perspective for both theory and practice is its compatibility with much empirical work, for data can be disaggregated concerning male and female individuals according to their own or others' categorisations. However, the conceptualisation of gender as a biological category of difference has been criticised as essentialist, assuming a biological gender identity and homogeneity of "men" and "women" as collective groups that does not exist.

Societal category

Gender as a societal category tries to overcome the individual and essentialist perspective while emphasising the social construction and societal embeddedness of gender. Consequently, gender perspectives address gender relations in their historical, societal and cultural context. While some European languages differentiate between a biological sex (Spanish: sexo, French: sexe) and a social gender (Spanish: género, French: genre) others do not (e.g. German or Finnish) and therefore use the English term gender to signify the latter. Within the variety of gender approaches the question of how far materiality (still) exists – for example with regard to a biological sex – is crucial and heavily discussed (see for old and new feminist materialism Bauhardt, 2017). Moreover, one may differentiate between perspectives focusing on gender as a structural category and those which address the process of (un)doing gender.

Structural category

Gender as a structural category focuses on the configurations and conditions leading to the appreciation and devaluation of gendered fields of work, social positions, etc. (Aulenbacher, 2008). Critique of the separation and hierarchisation of a productive (male) and a reproductive (female) sphere outlines a key discussion within this understanding. In addition, there are other dichotomous patterns which are directly linked to the gendered separation of production versus reproduction (e.g. public versus private, paid versus unpaid work). Furthermore, it is acknowledged that gender is one, but not the only, category structuring society. Research on intersectionality focuses on the nexus of these societal structural categories (Crenshaw, 1989; Knapp, 2005; see below), and debates on "gender and diversity" or "gender+" broaden the perspective with regard to categories such as "race", "class", "age", etc.

Process category

Another shift within gender studies is marked by an understanding of gender as a process category. Based on the assumption that neither sex nor gender are fixed categories, gender as a process category opens up a perspective which inquires into the interactive (co)production of gender, "doing gender" (West and Zimmerman, 1987; Butler, 2004; Gildemeister, 2008). It can be linked to approaches focusing on the dimensions of *gender identity*, *gender expression* and (*sexual) orientation*, their interaction and disruption. From a poststructuralist point of view, *queer* theories not only address the construction of (gender or sexual) identities, but are also interested in "how norms and categories are deployed" (Oswin, 2008: 96). From this point of view, they criticise standardisation and homogenisation (Engel, 2005). This kind of looking for the in-betweens (e.g. naturecultures (Haraway, 2003), (re)productivity (Biesecker and Hofmeister, 2010)) seems to be inspiring for theoretical reflections as well as for interventions and disruption in practice.

In sum, researchers and practitioners are confronted with a variety of gender concepts and perspectives. All of these perspectives are informed by particular understandings of the category gender that are not always properly distinguished or distinguishable. Current debates about gender and its implementation in theory and practice strive to consider its linkages to other categories of marginalisation and discrimination in order to prevent its becoming invisible as a key category of social analysis. In this ambiguous and tense field, debates on *diversity* are often criticised for diminishing the category gender to only one of many categories that need to be taken into account. As the diversity concept is frequently linked to business strategies aiming to maximise the economic success of companies, it may lose the critical intention of gender approaches that include a questioning of affirmative capitalist economic orientations. The feminist academic debate on *intersectionality* makes it clear that such a limited gender perspective has to be overcome in order to look critically at relations of power and domination. By investigating intersections of different categories of marginalisation and discrimination, intersectionality addresses societal structures as a whole, but remains strongly linked to women's and gender studies where it has become a new paradigm (Winker and Degele, 2009). It is important in all these debates to remember that not only gender, but also categories such as "race" or "class" must not be understood in essentialist ways, rather their social construction needs to be constantly considered. In order to take up these discussions and to underline the crucial significance that the category gender still has, new concepts such as *gender+* (Alsos, Ljungren and Hytti, 2013) aim to conceptualise gender as a social category complemented by and linked to these other socially constructed categories.

Space

Space is another important concept in the context of this volume, and can simultaneously serve as another epistemological category. In spatial planning and spatial sciences, it has been – and continues to be – conceptualised in manifold ways. Spatial concepts used in geography, planning theory and practice have undergone a transformation over time – from a positivist mode to relational and social-constructivist approaches (see e.g. Läpple, 1991; Gottdiener, 1994; Sturm, 2000; Massey, 2005; Davoudi, 2009, 2012). Anglo-American or German planning contexts are arguably more inflected by these concepts than in countries such as France, Greece and Spain in which spatial concepts have been particularly influenced by urban design, urbanism and architecture debates. At the same time, however, authors with a critical social science background such as Henri Lefebvre and Manuel Castells have been greatly influential for planning debates in these countries, too.

The so-called "spatial turn" of the 1980s (Bachmann-Medik, 2016) drew attention to space as a central category for social and cultural sciences. For example, the work of the French sociologist Henri Lefebvre (1974/1991) introduced a threefold understanding of space in physical, mental and social terms: le perçu, le conçu et le vécu, in English: "the perceived, the conceived, and the lived" (Lefebvre 1974/1991: 220). In his model, space is not an abstract container which may be filled up with human ideas, concepts, buildings and settlements, but a category exploring the relations between physical-material, psychological-individual and social-common elements, leading to differentiated as well as changeable interpretations (see Figure 1.1).

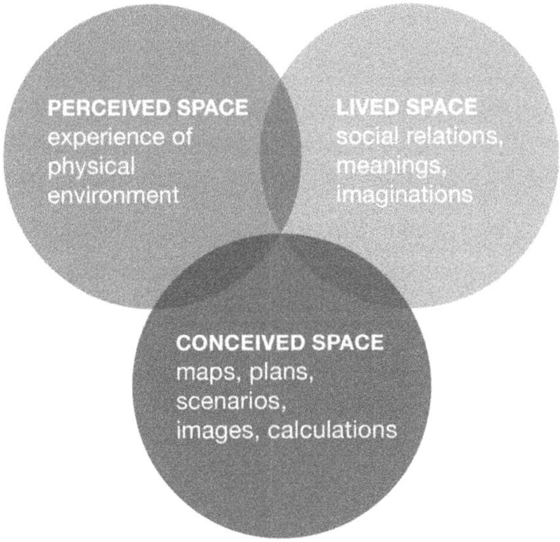

Figure 1.1 Spatial dimensions according to Henri Lefebvre (1974/1991).
Source: © Huning, Mölders and Zibell.

Lefebvre's concept highlights the strong interrelations between space as structure and space as social process shaped by individuals' conceptualisations, perceptions and everyday lives. Thus, spaces are the result of syntheses of perceived facts and interpretations; they are never finished or complete, but are permanently updated by new social and material interventions as well as by new representations. Relational concepts of space such as Lefebvre's do not confine themselves to temporal change in a given territory understood in the sense of a physical space with clear boundaries; rather, they draw attention to the simultaneity of spaces which overlap in one particular place and overcome borders. This simultaneity results in distinct perceptions, needs and options of appropriation and use for different social groups and genders. Understanding space "as the product of interrelations" (Massey, 2005: 9), British geographer Doreen Massey saw space as "a cut through the myriad stories in which we are all living at any one moment" (Social Science Bites, 2013: 3) and a sphere of "contemporaneous plurality" (Massey, 2005: 9). She stressed the simultaneity and multiplicity of experiences in different places at single points in time, even in one single place. In this reading, space is a mirror, an object and expression of societal power relations and of more or less obvious power struggles. As a consequence, space is always "under construction" (Massey, 2005: 9), potentially open to the future, and contingent. In this vein, concepts like "performing space" or "doing space" (Rose, 1999; Strüver, 2018) are important leads for understanding the role of planning for spatial development. In particular when it comes to interrelations with gender, relational and thus dynamic conceptions are indispensable because they allow a non-essentialist understanding of gender and space to be taken into account not only in structural but also in procedural terms.

British planner Simin Davoudi argues that over past decades, the dominant understanding in international (mostly Anglo-Saxon) planning debates has changed from a positivist to a more interpretative tradition: "Fluidity, contingency, dynamism and simultaneity are key characteristics of interpretive planning" (Davoudi, 2012: 438). This also has consequences for the way in which space is conceptualised (see Table 1.2).

Relational conceptions of space are part of the interpretative tradition in planning. They allow planners to better understand the variety of (e.g. gender) perspectives which may come together in a single place and that planning may be more complicated than designing a territory for a common good with mostly rational means. But for planners in research and particularly in practice, relational concepts of space are difficult to handle. Davoudi speculates that "the interpretive approach will find it difficult to offer planners alternative ways of articulating spatiality in the face of the renewed influence of positivism" (Davoudi, 2012: 439), which she and other authors detect in reactions to new global conflicts and environmental challenges. As a consequence, the social construction of space which lies at the heart of many relational concepts of space is as difficult to mainstream in planning as is a social-constructivist understanding of gender, although these post-positivist, interpretative approaches seem particularly suited to meet the above-mentioned need to integrate economic, social and environmental aspects in pursuit of sustainable development. They also support critical thinking about spatial development and the deconstruction of underlying ideologies, assumptions and taken-for-granted presuppositions.

A relational perspective is also helpful for describing, measuring and labelling spatial differentiation (Huning, Bens and Hüttl, 2012). All descriptions, measures and labels can be understood as social constructions, and are quite

Table 1.2 Transformation from a positivist to an interpretative planning tradition

	Positivist tradition	*Interpretative tradition*
Spatial conception	Neutral container, objectively measurable	Relational space
Scale and order	Control of socio-spatial complexity	Recognition of complexity, search for potentials
Time, future	Linear, reduction of insecurities	Fluid, open search for current opportunities
Planning process and plans	Linear, rational, result of solid databases	Result of creativity, learning, storytelling
Representation, visualisation	Maps	Scenarios, "fuzzy maps"

Source: adapted from Davoudi, 2012.

powerful in organising spatial relations. Supported by political and administrative institutions, regulatory frameworks, collective narratives, etc., they enter collective spatial imaginaries embedded in material spatial structures and thus physical spaces. One important case in point is scale. Scales are the outcomes of socio-spatial dynamics, as they "are strategic, contingent and politically powerful concretisations of fundamentally fluid social processes" (McCann, 2004: 2319; also see Swyngedouw, 2004). At the same time, they structure everyday life, organise citizens' rights and their socio-economic inclusion and exclusion, and generate feelings of belonging and identity. Geographers Sallie Marston, John Paul Jones III and Keith Woodward propose "abandon[ing] hierarchical scale in its entirety" (2005: 420) in favour of a "flat ontology concerned with both the world's very real potentialities and actualities" (2005: 426). While such a multiscalar and non-hierarchical analysis corresponds to the need to integrate the different dimensions of sustainability – social, economic and environmental – and the different spatial – and consequently political and administrative – levels, it is likely that this extended level of complexity may exceed the coping capacity of spatial policy and planning.

In relation to gender, we can see that social constructivist approaches inspire thinking about gender equality and equity in terms of other categories of social differentiation, as they inspire thinking about physical and socio-spatial relations. However, as Lidewij Tummers, Sylvette Denèfle and Heidrun Wankiewicz argue in Chapter 3, resistance to these more complex conceptualisations of space and gender – which is reflected, for example, in guidelines that reduce gender mainstreaming to a merely technical operation with regard to housing, transport or open space design – may be related to general (conscious or subconscious) hegemonic thinking which is not able (nor particularly willing) to actually question the current state of the art (also see Sánchez de Madariaga, 2013).

Space between development and planning

Space offers the physical and social framework for every human existence and action. Settlements are located and built within "given" historical structures that emerge over time, landscapes are shaped and transformed over and over again. In this process, the answer to the question of who is able or allowed to act as a decision-maker and who has to accept the decisions others have made is a matter of culture and society, of social, political and economic hierarchies and gender relations. At the same time, every human action concerning space is affected by change, for example in terms of demography or environmental conditions such as climate change. Thus, spatial development is a result of various intended and unintended changes.

In transformation processes, space is conceptualised in different ways depending on whether it is looked at as an object of architecture and urban design/urbanism or as an object of spatial planning and territorial development. At the same time, as described above (see the second section), different scales lead to different perspectives on space. Understandings of spatial development and planning, therefore, vary from two- and three-dimensional to four-dimensional approaches, being more or less figurative or abstract. Last but not least, national context, language and political culture influence which comprehension of space and spatial development is pre-dominant in a particular setting.

In many parts of Europe, (urban) planning was "invented" in order to organise spatial development when spatial relations became more and more complex. It emerged during the course of industrialisation in response to rapidly growing towns and the emergence of metropolitan areas from the end of the nineteenth century. Established as a public institution in countries such as Austria, Germany and the UK especially since the post-World War II period, its mission was to regulate spatial development to ensure human welfare and provide common goods, and thereby to balance socio-spatial disparities and inequalities. Spatial "planning" was understood and treated as a counterpart to economic "development", state as opposed to market. This understanding was transformed when public budgets declined during the 1980s, and new public management and new governance strategies – such as public-private partnerships – became essential to provide infrastructures or to reorganise former industrial sites in urban areas. Planning – hitherto mostly conceived and executed as top-down, by public administrations or as an extended branch of public administration – turned into a more "cooperative" field where different political and economic stakeholders worked together to shape spatial development. Development was no longer only an economic concept but found its way into the field of urban (and regional) or spatial planning and decision-making, as a field of action and an academic discipline. In the northern parts of Europe, it became a professional field distinct from architecture (and urbanism). New forms of cooperative planning at different scales also appeared due to merging settlement areas in growing urban territories or because of shared service provision especially in rural regions. From the mid-1990s on, the European Union funding system helped to establish a new understanding of "regional development" with tools such as the European Social Fund (ESF), the European Regional Development Fund (ERDF) or the European Agricultural Fund for Rural Development (EAFRD). The LEADER initiative was a good case in point since the installation of new regional forms of cooperation and thus new governance structures were its explicit goals. Via these funding programmes, the gender mainstreaming approach could disseminate across Europe (see below). In contrast, in other European countries such as France (including French- or Italian-speaking Switzerland), Greece and Spain, planning is strongly linked to architecture and urbanism both in theory and practice. This results in approaches to spatial development that are more design- than (only) process-oriented. Accordingly, the options for integrating and implementing gender (mainstreaming) perspectives into spatial policies and planning are different from those mentioned before.

The term "spatial development" refers to evident changes of space in physical and social terms resulting from economic and political decisions and with more or less formal shareholder and citizen participation (depending on the particular political system). In this way, it is both passive, affected by many factors as mentioned above, and active, intentionally pursued by planners, architects and developers (see Wallin, 2018). In professional contexts, spatial development as a term is often related to the actions of institutional or economic stakeholders, but in fact it also takes place in everyday life performed by everybody – of all genders – who uses space and moves in space.

"Spatial planning" in contrast refers to the ways in which states, countries and municipalities intentionally rule, organise and manage their spatial development, preparing decisions on building and settlement consolidation and extension. It is a small institutional field responsible for regulating processes that guide space and landscapes towards common welfare. In recent decades it has changed from an exclusively top-down approach, as a responsibility of public administrations, and now includes more and more cooperative forms in which people are, or are at least intended to be, more involved. As a consequence, (non-formal) bottom-up activities have become much more important than they used to be. In the course of this debate, various definitions and conceptions of space ranging from mere physical space to "an unfinished social project" have also found their way into planning practices whose task is to manage people's coexistence in the shared spaces of cities and neighbourhoods, aiming to enrich human life and to work towards social, cultural and environmental justice.

Planning cultures and policies controlling spatial development differ in the various European countries (as reflected in this volume), depending not least on time and space. For this reason, we use the term "spatial development" instead of "spatial planning" as a broader perspective on transformation processes: the term development allows room for new ways of decision-making and planning innovation. It offers possibilities for critical perspectives and flexibility between top-down and bottom-up approaches in favour of more cooperative settings on an equal footing. This may also provide a chance for feminist approaches which traditionally do not emerge out of set formal or institutional structures. At the same time, of course, it also implies risks because of new forms of governance which cede more power to economically powerful stakeholders.

Gender in spatial development

Gendered perspectives in spatial development are not at all common in planning practice today. If they are applied, this is mostly the result of a favourable constellation made up of open-minded chief officers for urban planning and development together with equal opportunity commissioners (Grüger and Zibell, 2005). They are also established in research projects and programmes, funded by applied research institutions (for the German case, see e.g. BBR, 2006) or by the European Union (e.g. GenderAlp!; EU Community Initiative Alpine Space Programme, 2005–2007), mostly as pilot projects or one-off interventions, seldom as attempts to carry out a continuous integration and implementation process (though see the cities of Vienna and Munich for exceptions).

In planning theories, gendered perspectives are likewise rarely part of the mainstream and remain very much on the sidelines (cf. Zibell, 2012; Damyanovic and Zibell, 2013), although recently, they have surfaced in the context of debates about a "just city" (Marcuse *et al.*, 2009; Fainstein, 2010). Such debates tackle the question of justice – particularly in terms of democracy, diversity and equity (understood in the sense of "fairness") in socio-spatial settings.

To evaluate if and how the category "gender" is considered in the context of spatial development it is necessary to reflect on the underlying concepts of "space" and "gender" within different planning approaches (Huning, 2013). For this purpose, the analytic gender perspectives presented here offer a helpful classification and will be referred to below, in regard to the respective spatial concepts.

Linking gender concepts with spatial development and planning

The understanding of *gender as a category of difference* leads to what is known as women-oriented or women-adequate planning (e.g. avoidance of places of fear). In the post-World War II environment, there was a search for the re-valuation of the female and for a stronger recognition of women as political and social subjects. In order to address women's everyday experiences, needs and realities, women were identified as a specific "target group" with specific spatial needs. This incidentally increased the visibility of women and was at the same time a first step away from standardising people and towards recognising the need for differentiation. In consequence, the awareness for sex- and gender-differentiated statistics and data collection increased, reflected by an incremental release of gender atlases edited by towns and regions, states and also the European Union (e.g. Switzerland: Bühler, 2001; Germany: BBR, 2007; BBSR, 2014; EU: EIGE, 2017). However, this target-oriented approach was criticised, not only by feminist planning theorists, as being essentialist and neglecting the heterogeneity of women and the social construction of the category gender. To this day, this type of essentialisation of gender often correlates with what was introduced as the conception of space as a container. Thus, the critique of neglecting the social sphere was applicable to both categories in their specific interrelation. Continuous efforts were therefore made to move towards more societally based space and gender approaches.

The understanding of *gender as a structural category* tries to overcome the individual and essentialist perspective of difference (more or less successfully) while addressing gender as a societal category, also in development and planning (e.g. Berlin: Senate Department for Urban Development, 2011; Vienna: Damyanovic, Reinwald and Weikmann, 2013). Through this approach, the structural conditions leading to the appreciation and devaluation of gendered fields of work are brought into focus, rather than the individuals who are doing this work – no matter which gender. An example of this for planning would be to consider or even privilege the requirements of domestic work while spatialising and designing housing settlements and facilities. However, this understanding of gender is not yet grasped or implemented in most spatial development and planning (see Chapters 4 and 5). Also, in the context of gender mainstreaming (see below) the planning officers responsible often fall into the "gender trap" whereby the differentiation between women and men and the old corresponding allocations of what is associated with male or female repeatedly creeps in through the backdoor (see Chapter 3).

Understanding *gender as a process category* in planning might lead to more sensitivity regarding the connection between doing space and doing gender. The shift in gender studies towards a social constructivist perspective has already – at least to some extent – narrowed the gap between gender studies and spatial planning in this direction. At the same time, the call for "deconstruction" was perceived by many planners as merely being academic debate with no relevance for everyday professional practice. Research has shown how gender, space and sexuality are co-constructed in many regards (Frisch, 2002; Schuster, 2010) and this calls for the need to "queer" spaces (Hark, 2004; Oswin, 2008), but how these insights can be transferred to normative planning is still unclear. Planning has taken advantage of more relational and performative approaches, particularly where towns and cities are in economic decline and there is a depreciation of property values (Tornaghi and Knierbein, 2015), and also in terms of participation (Turnhout, van Bommel and Aarts, 2010). But this shift is usually more related to the interpretative turn in planning (see above) than to an acknowledgement of the relevance of gender. A combination of theoretical perspectives from gender studies and recent (performative) relational planning approaches might support a wider understanding of planning that would be able to open up spaces for both "deconstructing" and "doing" space.

Gender in spatial development stands for a gender-aware and gender-sensitive attitude. In this way, it addresses every institution and every stakeholder who deals with spatially relevant decisions, actions and measures. To further such an attitude among planning and developing stakeholders, political will and legal regulations are essential. The following table – an adapted version of the table introduced in the gender section above (Table 1.1, Page 4) – shows the way in which the gender categories, their meanings and related methodologies could be systematically applied to planning issues (see Table 1.3).

Table 1.3 Analytic categories of gender with related methodologies and planning issues

Category	*Issues for planning*
Biological category (Sex)	Women-only housing; women's shelters, housing for single parents/mothers; women's representation in planning/participation processes, etc.
Societal categories	
• Structural category (Gender)	Integration of reproduction and production spheres at different scales; revision of image production, overcoming gender stereotypes; partial dissolution of the public-private binary, etc.
• Process category (Queer, gender identity, expression, orientation)	Opportunities for (un)doing gender+ in different situations and places at varying scales

The planning perspective often loses its gender focus when addressing such issues in practice. "We are planning for everyone", is the commitment of planners who feel dedicated to their duty of providing common welfare for everybody. In most cases there is an implicit attitude towards good planning in terms of user-oriented planning. Planners are convinced of the need to do the best for people and spatial development in general. But the same planners who are claiming such "good" planning for themselves often are those who refuse or do not understand why gender perspectives should be included in their plans and strategies.

The concept of diversity (planning) meets the commitment to planning for all in considering lots of (inner, outer and organisational) dimensions of difference (Gardenswartz and Rowe, 1995), gender being only one alongside others, without any hierarchisation or reference to power relations. In this way, diversity as well as "difference" in the UK has incorporated gender as one of many inequalities. With reference to Fincher and Iveson (2008), Roberts retraces three key categories which can assist in highlighting how subjectivities of difference interact with the formal planning system. The first category concerns the *distribution* of resources between groups, the second refers to the *recognition* of particular groups within democratic processes, and the third category – *encounter* – "takes in the opportunities that different groups in the population have to come upon each other" (Roberts, 2013: 7).

The process of consultation and the media and methods used in eliciting different viewpoints have also formed an important strand in the feminist approach in spatial planning. For example, Leonie Sandercock challenges epistemology in her arguments for changing the ways in which planners gather information. She stresses the importance of "listening and hearing", arguing that planners need to exercise empathy and imagination to understand how inequalities are experienced. This means becoming more inclusive in the methods used for consultation and moving beyond the conventional meetings and questionnaires to include storytelling and other narrative outputs (Sandercock and Forsyth, 1996; Sandercock, 2003; Roberts, 2013).

According to Damyanovic (2007) and Wankiewicz (2016), gender planning or gender-sensitive planning focuses on women and men and their relations as well as on gender-specific roles and stereotypes. It values spatial realities according to their qualities for (gendered) everyday life and includes age, period of life and social background. In this way it comes close to concepts of gender+ which accept the priority of gender realities and identities in simultaneously recognising lots of other differentiations (see intersectionality above). In uncovering societal conditions and values in planning, gender planning is transformative in that it aims to change unequal power relations and the planning concepts and instruments which underpin these (Damyanovic, 2007). An important element is participation and giving voice to all members of a society or neighbourhood (Wankiewicz, 2016) – particularly those whose voices tend to remain unheard. The aim of gender planning understood like this is to en-gender planning in an enduring and sustainable way (Horelli, 2017).

Gender mainstreaming in spatial development

En-gendering spatial development (Horelli, 2017) does not arise by itself, because systems are in many ways inert to change. Societal efforts and political will are essential preconditions for transformation towards gender justice and equity in spatial development. With the Treaty of Amsterdam (European Communities, 1997), the principle of Gender Mainstreaming (GM) was introduced at the level of the European Union (EU) and subsequently adopted by all EU member states. Thus, similarly to the concept of sustainability, it should be considered in every political field of action, as well as in spatial development and planning. There are indeed lots of attempts to implement both sustainable development and GM strategies, but their success is, to date, quite limited when basic transformations of society and socio-environmental relations are considered. Those measures, policies and projects that are actually applied tend to be adapted to current economic – capitalist – thinking; if they do not comply with this thinking there is little chance of them being implemented. Consequently, there are limited and scattered initiatives, here and there, but there has not yet been any serious integral and effective transformation towards gender justice and equity.

The significance of GM is that it is an organisational strategy to integrate a gender perspective to all aspects of an institution's policy and activities, through building gender capacity and accountability (Reeves and Baden, 2000). Gender planning, on the other hand, is defined as the technical and political processes and procedures necessary to implement gender-sensitive policy (Reeves and Baden, 2000), not specifically referring to spatial development. Nevertheless, among gender planning experts there is agreement about defining "gender planning" as "GM in spatial development" (Damyanovic, 2007; Wankiewicz, 2016).

As a political strategy, GM is formally adopted, and hence, still meaningful. A number of projects with a gender-sensitive approach have been funded with EU money (e.g. GenderAlp!, 2005–2007), leading to a stronger gender consciousness of all the stakeholders involved. The question of the extent to which and under what conditions these approaches are effective in terms of sustainable development has been investigated in the context of the aforementioned Interreg GenderAlp! project (Land Salzburg and Zibell, 2006). An essential insight was that the individual projects, once conceived, funded and carried out are not productive or sustainable unless they are underpinned by checks and evaluatory mechanisms as part of the process, and through embedding the learning to promote gender balance. Products, processes and structures have to come together in an inclusive strategy to be successful in mainstreaming gender in planning.

The gap between theory and practice or strategic needs on the one hand and practical needs on the other becomes obvious if we try to en-gender planning in practice (see also Chapters 2 and 3). Meeting the practical needs of people and supporting their everyday lives is of course much easier than meeting strategic gender needs (or interests) which challenge asymmetric power relations and aim to deconstruct and redefine the values of hegemonic masculinities (Damyanovic and Zibell, 2013). Planning first has to deal with what is there, in its respective political structures and under the influence of specific power relations. Without actors who – beyond applying rules and standards – have a vision of a different society, every en-gendering strategy is hard to implement. And the vision of a different society – e.g. in terms of a consistent reformation/definition of "good life" – and consequently of new ways of building, planning and designing space for different needs has to be constantly (re)negotiated in a permanent process of sustainable communication.

The extent to which the GM strategy has been successful in making development and planning or its outcomes more just and whether or not it has led to more equity between genders and along other intersectional lines are questions tackled by this publication. GM is only one – and a rather top-down – strategy, which should not be overestimated. With Roberts (2013: 12) we ask: "What comes after gender mainstreaming?" There are authors who argue that there are other, alternative or complementary ways to achieve gender equity and justice. For example, Horelli (2017) identifies four types of urban or spatial planning, distinguishing between top-down and bottom-up and between formal and informal planning practices and methodologies (see Figure 1.2).

However, it seems that none of these strategies are always and automatically effective and successful, and more gender-sensitive hybrid approaches to spatial development are needed.

Several thematic chapters in this volume demonstrate which concepts of and strategies for en-gendering space respectively cities or regions have been applied in different case studies, and comment on their respective successes and failures.

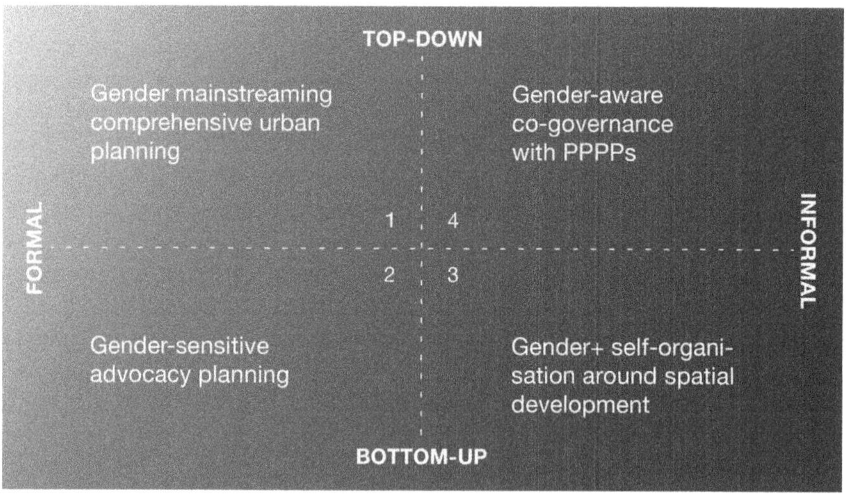

Figure 1.2 Different types of planning contexts require varying methods to en-gender spatial development according to Horelli (2017: 1781).

Source: © Horelli.

Sustainable spatial development – integrating concepts and debates

The 1990s saw an increasing relevance of topics such as environment, poverty and women's rights. Thus, issues which were and still are often marginalised and discussed differently in different parts of the world became part of the international political agenda and forced debates on social-ecological visions to face contemporary challenges. This is true for the United Nations Conference on Environment and Development (UNEP) held in Rio in 1992 as well as for the Fourth World Conference on Women held in Beijing in 1995. Whereas the first marked the beginning of intensive political and scientific debates on sustainable development, the second paved the way for gender mainstreaming as a political strategy.

The two movements did not take place separately, but with reference to one another. In particular, the sustainability discourse has been accompanied by an intensive articulation of women's rights and women's roles in realising sustainable development (Braidotti, 1999; Mölders, 2015; Wichterich, 2015). Feminist Non-Governmental Organisations (NGOs), such as the Women's Environment and Development Organisation (WEDO), Development Alternatives with Women for a New Era (DAWN) or Women in Development Europe (WIDE), formulate their own approaches to sustainability, emphasising the nexus between gender, societal relations to nature and economy. In this vein the Sustainable Livelihood approach of DAWN became a famous feminist vision of sustainability, one that is sceptical towards growth-oriented economic development (see Wiltshire, 1992; Wichterich, 2015). Within this process, women were acknowledged as a "major group" and the Agenda 21, as the key outcome of Rio, includes Chapter 24 "Global action for women towards sustainable and equitable development". But whereas the "spirit of Rio" was very much characterised by an integrative mode, the subsequent processes turned out to be more confrontational regarding gender perspectives. This is mainly due to the fact that the concepts and topics advocated by women's groups – such as sustainable livelihood, care, human rights and critiques of power – contradict the ongoing neoliberalisation of the sustainability discourse which culminated in the idea of a Green Economy (DCED and BMZ, 2012; Wichterich, 2015). Consequently, with regard to the Sustainable Development Goals (SDGs), the Women's Major Group (WMG) states: "instead of moving towards a green economy, the members of the WMG propose a transformation towards a sustainable, equitable and inclusive economy" (WMG, 2013: 13). In sum, it becomes obvious that sustainable development is a contested concept, connected with various contradictions and conflicting expectations. The broad debate on gender and sustainability opens up specific perspectives by criticising hegemonic rationalities and power relations, not only between all genders but also between different visions on how to create "our common future".

These findings on gender and sustainability, their lineages as well as their fractures, directly affect the topic of spatial development. In general, the conceptual framework of sustainability is shaped by two basic normative elements: the idea of intra- and intergenerational *justice* and the *integration* of environmental, economic and social spheres (Hofmeister, Mölders and Thiem, 2014). Justice and integration are directly linked to gender and affect spatial development from a feminist perspective. First, claiming intragenerational justice calls for the recognition of equal rights in the sense that equity is seen as one step towards equality. Second, the integration of spheres which are often treated as separate from each other – such as social, environmental and economic – is an implicit feminist concern and as such is able to connect and interrelate everyday life with space (and time). Feminist scholars criticise dualisms in thinking and action because of their gendered logic (e.g. the separation of paid, male work and unpaid, female work). This critique becomes vital in terms of spatial development, for separation finds its expression in spatial dimensions and leads to what is known as "a separation of functions", "the separation of private and public spaces" or "the separation of reproductive and productive spaces" (Terlinden, 1990), which is not realistic in practice.

In summing up, it has to be stated that especially from a gender perspective, sustainable and spatial development are inseparably linked. Thus, gender perspectives – taken seriously – not only help, but are downright necessary to promote transformative cultures towards sustainable (spatial) development.

Note

1 As Gottschlich and Bellina (2017) elucidate with reference to Shrader-Frechette (2002), equity is linked to a redistributive justice that takes into account and rectifies existing inequalities. It goes beyond justice in a more liberal sense "in which equality implies 'the same for every individual'" (Gottschlich and Bellina, 2017: 945; for gender equity and equality see also Reeves and Baden, 2000).

References

Alsos, G.A., Ljungren, E. and Hytti, U. (2013). Gender and innovation. State of the art and a research agenda. *International Journal of Gender and Entrepreneurship*, 5(3), 236–256.

Aulenbacher, B. (2008). Geschlecht als Strukturkategorie: Über den inneren Zusammenhang von moderner Gesellschaft und Geschlechterverhältnis (Gender as a structural category. On the interconnection of modern society and gender relations). In: S.M. Wilz (Ed.), *Geschlechterdifferenzen – Geschlechterdifferenzierungen. Ein Überblick über gesellschaftliche Entwicklungen und theoretische Positionen (Gender differences – gender differentiations. An overview of societal developments and theoretical positions)*. Wiesbaden: Springer VS, 139–166.

Bachmann-Medik, D. (2016). *Cultural turns. New orientations in the study of culture*. Berlin and Boston: De Gruyter.

Bauhardt, C. (2017). Living in a material world. Entwurf einer queer-feministischen Ökonomie (Living in a material world. A sketch for a queer-feminist economics). *Gender*, 9(1), 99–114.

BBR – Bundesamt für Bauwesen und Raumordnung (Ed.) (2007). *Frauen – Männer – Räume (Women – men – spaces)*. Bd. 26. Bonn.

BBR – Bundesamt für Bauwesen und Raumordnung (Ed.) (2006). *Gender Mainstreaming im Städtebau (Gender mainstreaming in urbanism)*. ExWoSt-Informationen, 26/5-04/2006. [Online] Bonn. Available at: www.bbsr.bund.de/BBSR/DE/Veroeffentlichungen/ExWoSt/24_29/exwost26.html [Accessed 23 May 2018].

BBSR – Bundesinstitut für Bau-, Stadt- und Raumforschung (2014). *Gender-Index*. [Online]. Available at: www.bbsr.bund.de/BBSR/DE/Raumbeobachtung/GenderIndex/GenderIndex_node.html [Accessed 23 May 2018].

Biesecker, A. and Hofmeister, S. (2010). Focus: (Re)productivity. Sustainable relations both between society and nature and between the genders. *Ecological Economics*, 69(8), 1703–1711.

Braidotti, R. (1999). Towards sustainable subjectivity. A view from feminist philosophy. In: E. Becker and T. Jahn (Eds.), *Sustainability and the social sciences. A cross-disciplinary approach to integrating environmental considerations into theoretical reorientation*. London and New York: Zed Books, 74–95.

Bühler, E. (2001). *Frauen- und Gleichstellungsatlas Schweiz (Gender equality atlas of Switzerland)*. Zürich: Seismo.

Butler, J. (2004). *Undoing gender*. New York and London: Routledge.

Crenshaw, K. (1989). Demarginalizing the intersection of race and sex. A black feminist critique of antidiscrimination doctrine, feminist theory and antiracist politics. *University of Chicago Legal Forum*, 1989(1), 139–167. [Online]. Available at: http://chicago unbound.uchicago.edu/uclf/vol. 1989/iss1/8 [Accessed 23 May 2018].

Damyanovic, D. (2007). *Landschaftsplanung als Qualitätssicherung zur Umsetzung der Strategie des Gender Mainstreaming (Landscape planning as quality assurance for the implementation of the strategy of gender mainstreaming)*. Vienna: Gutmann Perterson.

Damyanovic, D., Reinwald, F. and Weikmann, A. (2013). *Gender mainstreaming in urban planning and urban development*. Werkstattbericht, Nr. 130A. Vienna: Urban Development Vienna, Municipal Department 18 (MA 18) – Urban Development and Planning. [Online] Available at: www.wien.gv.at/stadtentwicklung/studien/pdf/b008358.pdf [Accessed 23 May 2018].

Damyanovic, D. and Zibell, B. (2013). Is there still gender on the agenda for spatial planning theories? Attempt to an integrative approach to generate gender-sensitive planning theories. *disP – The Planning Review*, 49(4), 25–36.

Davoudi, S. (2012). The legacy of positivism and the emergence of interpretive tradition in spatial planning. *Regional Studies*, 46(4), 429–441.

Davoudi, S. (2009). *Conceptions of space and place in strategic spatial planning*. London: Routledge.

DCED – The Donor Committee for Enterprise Development – and BMZ – Bundesministerium für wirtschaftliche Zusammenarbeit und Entwicklung (Eds.) (2012). *Women's participation in green growth – a potential fully realised*. Bonn and Berlin.

EIGE – European Institute for Gender Equality (Ed.) (2017). *Gender Equality Index 2017. Measuring gender equality in the European Union 2005–2015 – Report*. [Online]. Available at: http://eige.europa.eu/rdc/eige-publications/ [Accessed 23 May 2018].

Engel, A. (2005). Entschiedene Interventionen in der Unentscheidbarkeit. Von queerer Identitätskritik zur VerUneindeutigung als Methode (Definite interventions in undecided situations. From queer identity critique to producing ambiguity as a method). In: C. Harders, H. Kahlert and D. Schindler (Eds.), *Forschungsfeld Politik. Geschlechtskategoriale Einführung in die Sozialwissenschaften (Politics as research field. Gender categorical introduction to Social Sciences)*. Wiesbaden: Springer VS, 261–282.

EU Community Initiative Alpine Space Programme (2005–2007). *GenderAlp! Spatial development for women and men*. Land Salzburg – Planning, economic and equal opportunity department, Austria (Lead Partner). [Online]. Available at: www.alpine-space.org/2000-2006/genderalp.html [Accessed 23 May 2018].

European Communities (1997). *Treaty of Amsterdam amending the Treaty on European Union, the Treaties establishing the European Communities and certain related acts*. Luxembourg: Office for Official Publications of the European Communities.

Fainstein, S. (2010). *The just city*. Ithaca, NY: Cornell University Press.

Fincher, R. and Iveson, K. (2008). *Planning and diversity in the city. Redistribution, recognition and encounter*. Basingstoke: Palgrave Macmillan.

Frisch, M. (2002). Planning as a heterosexist project. *Journal of Planning Education and Research*, 21(3), 254–266.

Gardenswartz, L. and Rowe, A. (1995). *Diverse teams at work. Capitalizing on the power of diversity*. Alexandria, VA: Society for Human Resource Management.

Gildemeister, R. (2008). Soziale Konstruktionen von Geschlecht: “Doing gender” (Social constructions of gender: “Doing gender”). In: S.M. Wilz (Ed.), *Geschlechterdifferenzen – Geschlechterdifferenzierungen. Ein Überblick über gesellschaftliche Entwicklungen und theoretische Positionen (Gender differences – gender differentiations. An overview of societal developments and theoretical positions)*. Wiesbaden: Springer VS, 167–198.

Gottdiener, M. (1994). *The social production of urban space*. 2nd edn. Austin: University of Texas Press.

Gottschlich, D. and Bellina, L. (2017). Environmental justice and care. Critical emancipatory contributions to sustainability discourse. *Agriculture and Human Values*, 34(4), 941–953.

Grüger, C. and Zibell, B. (2005). Von der frauengerechten Stadtplanung zum Gender Mainstreaming in der Stadtentwicklung. Einblicke in die Planungspraxis (From women-based urban planning to gender mainstreaming in urban development. Insights from planning practice). In: Deutscher Städtetag (Ed.), *Frauen verändern ihre Stadt. Arbeitshilfe 4: Indikatoren und Gender Mainstreaming in der räumlichen Planung (Women change their city. Guideline 4. Indicators and gender mainstreaming in spatial planning)*. Köln: Deutscher Städtetag, 63–82.

Haraway, D. (2003). *The companion species manifesto. Dogs, people, and significant otherness*. Chicago: Prickly Paradigm.

Haraway, D. (1988). Situated knowledges. The science question in feminism and the privilege of partial perspective. In: *Feminist Studies*, Bd. 14, No. 3, 575–599. [Online]. Available at: www.jstor.org/stable/3178066 [Accessed 1 Nov. 2018].

Harding, S. (1991). *Whose science? Whose knowledge? Thinking from women's lives*. Ithaca, NY: Cornell University Press.

Hark, S. (2004). “We're here, we're queer, and we're not going shopping!” Queering space. Interventionen im Raum (“We're here, we're queer, and we're not going shopping!” Queering space. Interventions in space). In: C. Bauhardt (Ed.), *Räume der Emanzipation (Spaces of emancipation)*. Wiesbaden: Springer VS and GWV Fachverlage GmbH, 221–234.

Hofmeister, S. and Katz, C. (2011). Naturverhältnisse. Geschlechterverhältnisse. Nachhaltigkeit (Nature relations. Gender relations. Sustainability). In: M. Groß (Ed.), *Handbuch Umweltsoziologie (Handbook of environmental sociology)*. Wiesbaden: Springer VS, 365–398.

Hofmeister, S., Katz, C. and Mölders, T. (2013). Grundlegungen im Themenfeld Geschlechterverhältnisse und Nachhaltigkeit (Foundations in the thematic field of gender relations and sustainability). In: S. Hofmeister, C. Katz and T. Mölders (Eds.), *Geschlechterverhältnisse und Nachhaltigkeit. Die Kategorie Geschlecht in den Nachhaltigkeitswissenschaften (Gender relations and sustainability. The category gender in Sustainability Sciences)*. Opladen, Berlin and Toronto: Verlag Barbara Budrich, 33–76.

Hofmeister, S., Mölders, T. and Thiem, A. (2014). Nachhaltige Raumentwicklung (Sustainable spatial development). In: H. Heinrichs and G. Michelsen (Eds.), *Nachhaltigkeitswissenschaften (Sustainability sciences)*. Berlin and Heidelberg: Springer Verlag, 523–547.

Horelli, L. (2017). Engendering urban planning in different contexts – successes, constraints and consequences. *European Planning Studies*, 25(10), 1779–1796.

Horelli, L. and Wallin, S. (2013). Gender-sensitive e-planning for sustaining everyday life. In: I. Sánchez de Madariaga and M. Roberts (Eds.), *Fair shared cities. The impact of gender planning in Europe*. Farnham: Ashgate, 330–353.

Huning, S. (2013). *Ambivalences of gender planning*. [Online]. Available at: www.metropolitiq ues.eu/Ambivalences-of-Gender-Planning.html [Accessed 23 May 2018].

Huning, S., Bens, O. and Hüttl, R.F. (2012). Demographic change beyond the urban-rural divide. Re-framing spatial differentiation in the context of migration flows and social networks. *Die Erde*, 143(1–2), 153–172.

Keller, E.F. (1995). The origin, history and politics of the subject called "gender and science". A first person account. In: S. Jasanoff, G.E. Markle, J.C. Petersen and T. Pinch (Eds.), *Handbook of science and technology studies*. Thousand Oaks, CA, London and New Delhi: Sage, 80–94.

Knapp, G.-A. (2005). "Intersectionality" – ein neues Paradigma feministischer Theorie? Zur transatlantischen Reise von "Race, Class, Gender" ("Intersectionality" – a new paradigm for feminist theory? About the transatlantic journey of "race, class, gender"). *Feministische Studien*, 23(1), 68–81.

Land Salzburg (Ed.) and Zibell, B. (2006). *Requirement-oriented spatial planning. Gender practice and criteria in spatial planning, final report*. English Abstract, Materialien zur Raumplanung, Bd. 22, Salzburg.

Läpple, D. (1991). Essay über den Raum (Essay on space). In: H. Häußermann, D. Ipsen and T. Krämer-Badoni (Eds.), *Stadt und Raum. Soziologische Analysen (City and space. Sociological analyses)*. Pfaffenweiler: Centaurus, 167–183.

Lefebvre, H. (1974/1991). *The production of space*. Translation of *La production de l'espace*. Malden, MA, Oxford and Victoria: Blackwell Publishing.

Marcuse, P., Connolly, J., Novy, J., Olivo, I., Potter, C. and Steil, J. (Eds.) (2009). *Searching for the just city. Debates in urban theory and practice*. London and New York: Routledge.

Marston, S.A., Jones III, J.P. and Woodward, K. (2005). Human geography without scale. *Transactions of the Institute of British Geographers*, 30(4), 416–432.

Massey, D. (2005). *For space*. London, Thousand Oaks, CA, New Delhi and Singapore: Sage.

McCann, E.J. (2004). Urban political economy beyond the "global city". *Urban Studies*, 41(12), 2315–2333.

Mölders, T. (2015). Geschlechterverhältnisse und Nachhaltigkeit – politische Strategien und wissenschaftliche Konzepte (Gender relations and sustainability – political strategies and scientific concepts). In: K. Bueschges (Ed.), *Bildung – Selbst(bild) – Geschlechterbilder (Education – self(image) – gender images)*. Berlin: Lit Verlag, 39–63.

Oswin, N. (2008). Critical geographies and the uses of sexuality. Deconstructing queer space. *Progress in Human Geography*, 32(1), 89–103.

Reeves, H. and Baden, S. (2000). *Gender and development. Concepts and definitions. Prepared for the Department for International Development (DFID) for its gender mainstreaming intranet resource.* BRIDGE (development – gender), report no. 55, Brighton: Institute of Development Studies.

Roberts, M. (2013). Introduction. Concepts, themes and issues in a gendered approach to planning. In: I. Sánchez de Madariaga and M. Roberts (Eds.), *Fair shared cities. The impact of gender planning in Europe*. Farnham: Ashgate, 1–18.

Rose, G. (1999). Performing space. In: D. Massey, J. Allen and P. Sarre (Eds.), *Human geography today*. Cambridge: Polity, 247–259.

Sánchez de Madariaga, I. (2013). Looking forward, moving beyond trade-offs. In: I. Sánchez de Madariaga and M. Roberts (Eds.), *Fair shared cities. The impact of gender planning in Europe*. Farnham: Ashgate, 325–333.

Sandercock, L. (2003). Out of the closet. The importance of stories and storytelling in planning practice. *Planning Theory and Practice*, 4(1), 11–28.

Sandercock, L. and Forsyth, A. (1996). Feminist theory and planning theory. The epistemological linkages. In: S. Campbell and S. Fainstein (Eds.), *Readings in planning theory*. Oxford: Blackwell, 471–474.

Schuster, N. (2010). *Andere Räume. Soziale Praktiken der Raumproduktion von Drag Kings und Transgender (Other spaces. Social practices of the production of space by drag kings and transgender)*. Bielefeld: transcript Verlag.

Senate Department for Urban Development Berlin (Ed.) (2011). *Gender in mainstreaming urban development. Berlin on the path towards becoming a metropolis worth living in for women and men.* [Online] Berlin. Available at: www.stadtentwicklung.berlin.de/soziale_stadt/gender_mainstreaming/download/gender_broschuere_englisch.pdf [Accessed 23 May 2018].

Shrader-Frechette, K. (2002). *Environmental justice. Creating equity, reclaiming democracy*. Oxford: Oxford University Press.

Social Science Bites (2013). Doreen Massey on space [podcast]. *Social Science Bites: A new podcast series with leading social scientists*. Available at: www.socialsciencespace.com/2013/02/podcastdoreen-massey-on-space/ [Accessed 23 May 2018].

Strüver, A. (2018). Performing spaces – gender relations, corpo-realities and space. In: C. Onnen and S. Rode-Breymann (Eds.), *Wiederherstellen – Unterbrechen – Verändern? Politiken der (Re-)Produktion. (Restoring – interrupting – changing? Policies of (re-)production)*. L'AGENda Bd. 3. Opladen, Berlin, Toronto: Verlag Barbara Budrich, 79–91.

Sturm, G. (2000). *Wege zum Raum. Methodologische Annäherungen an ein Basiskonzept raumbezogener Wissenschaft (Pathways to space. Methodological approaches to a basic concept of space-related science)*. Opladen: Leske Verlag und Verlag Barbara Budrich.

Swyngedouw, E. (2004). Globalisation or "glocalisation"? Networks, territories and rescaling. *Cambridge Review of International Affairs*, 17(1), 25–48.

Terlinden, U. (1990). *Gebrauchswirtschaft und Raumstruktur. Ein feministischer Ansatz in der soziologischen Stadtforschung (Use economy and spatial structure. A feminist approach in sociological urban studies)*. Berlin: Silberburg-Verlag.

Tornaghi, C. and Knierbein, S. (Eds.) (2015). *Public space and relational perspectives. New challenges for architecture and planning.* London: Routledge, 148–166.

Turnhout, E., van Bommel, S. and Aarts, N. (2010). How participation creates citizens. Participatory governance as performative practice. *Ecology and Society*, 15(4), 26. [Online]. Available at: www.ecologyandsociety.org/vol. 15/iss4/art26www.ecologyandsociety.org/vol. 15/iss4/art26. [Accessed 25 May 2018].

Wallin, S. (2018). *Managing urban complexity. Participatory planning, self-organisation and co-production of space*. Dissertation. Aalto University Espoo.

Wankiewicz, H. (2016). *Gender Planning – Gender Mainstreaming in der räumlichen Planung. Top-down- und Bottom-up-Strategien als Bausteine zu einer nutzerInnen- und gleichstellungsorientierten feministischen Raumplanung (Gender planning – gender mainstreaming in spatial planning. Top-down and bottom-up strategies as part of a user and gender equality-oriented feminist spatial planning)*. Dissertation. Paris Lodron Universität Salzburg.

West, C. and Zimmerman, D.H. (1987). Doing gender. *Gender and Society*, 1(2), 125–151.

Wichterich, C. (2015). Livelihood, Ökonomisierung des Lebendigen und gutes Leben – Feministische Perspektiven auf die verschlungenen Wege von Rio 1992 nach Rio 2012 (Livelihood, economisation of the living and a good life – feminist perspectives on the intertwined paths from Rio 1992 to Rio 2012). In: C. Katz, S. Heilmann, A. Thiem, K. Moths, L.M. Koch and S. Hofmeister (Eds.), *Nachhaltigkeit anders denken. Veränderungspotenziale durch Geschlechterperspektiven (Thinking sustainability differently. Potential of change through gender perspectives)*. Wiesbaden: Springer VS, 31–42.

Wiltshire, R. (1992). *Environment and development. Grass roots women's perspective*. Kingston: Dawn.

Winker, G. and Degele, N. (2009). *Intersektionalität. Zur Analyse sozialer Ungleichheiten (Intersectionality. On the analysis of social inequalities)*. Bielefeld: transcript Verlag.

WMG – Women's Major Group (2013). *Gender equality, women's rights and women's priorities. Recommendations for the proposed Sustainable Development Goals (SDGs) and the Post-2015 Development Agenda*.

Zibell, B. (2012). Planungstheorie und Gender. Geschlechterverhältnisse und Androzentrismen (Planning theory and gender. Gender relations and androcentrisms). In: R. May and B. Zibell (Eds.), *GenderKompetenz in Architektur Landschaft Planung. Ideen – Impulse – Initiativen (Gender competence in architecture landscape planning. Ideas – impulses – initiatives)*. Hannover: Internationalismus-Verlag, 84–86.

2 International and national perspectives

Similarities and differences

The main objective of this chapter is to tie together the history and future of gendered approaches in spatial development and planning in Europe on different scales and levels. The chapter begins with a brief historical review of spatial development and planning in general. In this section the authors first introduce different gendered approaches from the perspective of German-speaking countries from the nineteenth to the twenty-first century. The following section gives an overview of the planning systems and gendered approaches of nine European countries: Austria, England, Finland, France, Germany, Greece, the Netherlands, Spain, and Switzerland. The chapter concludes with insights from a comparison of the nine countries.

Brief historical review of gendered approaches in spatial development and planning

Doris Damyanovic and Barbara Zibell

Spatial development – how it "all" began…

Spatial development as defined in Chapter 1 is an expression of processes of "natural" and social change as well as human transformation. The construction of towns has, since ancient times, been considered as a cultural act shaping space and society. The planned construction of towns and cities in Europe can be traced back to the first town plan by Hippodamos of Miletus around 432 BC (Benevolo, 1983). Political power and responsibility for the architectural form of towns generally went hand in hand until well into medieval times and the early modern period. At any time, the type of society, political and economic organisations and power relationships moulded the drivers of change. For millennia, patriarchy was the supreme standard for European countries (and others) and therefore characterises the identity of change and transformation.

"Feminist" perspectives calling into question the prevailing (patriarchal) balance of power can be traced back at least to the High Middle Ages (see Zibell, 2013). An early work by French female author and philosopher Christine de Pizan appeared in 1405 (Pizan, 1405). Evidence of buildings of the Beguines was found along the River Rhine in areas of Germany, Belgium and the Netherlands, with built traces still visible today. Espousing class-independent forms of living, intellectual autonomy and economic independence (Triest, 1998; Reichstein, 2001; Simons, 2001; Unger, 2005), the Beguines lived in their own structural and social spatial units, known as "beguinages", which made a confident appearance within the shaped urban landscape. In various places, women as nuns and members of convents were also regularly involved in construction processes until the twentieth century (e.g. Busch-Sperveslage, 1999).

With the far-reaching economic and social changes of the industrial revolution during the nineteenth century (population growth, migration to the towns) the planned control of urban growth became a central task for the emerging public administrations. Corresponding plans and development concepts can be traced back to the middle of the nineteenth century in most European countries. In Germany regulation began to be enforced, restricting the freedom to build even as early as the end of the eighteenth century in order to avert public hazards, ensure safety and control impact on the landscape. Initial building regulations were issued for the first time in 1833 in Greece (Albers, 1997[1]). The United Kingdom and the Netherlands followed with laws focusing on the sanitation problems and hygiene of general urban expansion. In most countries, municipal planning skills began to be gradually introduced, together with the first development concepts: in 1834 in Athens, in 1855 in Barcelona and in 1862 in Berlin. The objective of these concepts was the orderly development of urban growth. Women's voices joined in the urban development debate on housing shortages, healthcare and urban hygiene, but could barely be heard against the concert of a male-dominated specialist discourse. The first comprehensive work on urban planning in Germany – a real theory of the architecture of large towns and cities on behalf of health and well-being – was published under the pseudonym Arminius by Countess Adelheid von Dohna-Poninska (1874) in Berlin (Albers, 1997). The text received insufficient recognition alongside three much better-known publications by Sitte (1889), Stübben (1890) and Baumeister (1876), and it was not until very recently that it began to attract more attention thanks to the heightened prestige and reprints of the works of feminist academics (Buchmüller, 1995; Frey and Perotti, 2015). Relevant reference works on town construction and town planning (a term already in common use in the UK; see Unwin, 1909) also started to emerge in other European countries, e.g. Spain, France and the Netherlands (Albers, 1997).

Criticism of urban development in the nineteenth century led to the introduction of special urban planning lectures (TU Vienna 1901, Berlin 1908) and gradually also to the establishment of the first urban development professorships at architectural faculties (University of Liverpool 1909, TU Delft 1924). Urban construction and urban planning laws followed in the first decades of the twentieth century (1909 UK, 1919 F, 1923 GR, 1931 FIN). Academic training in urban planning was introduced in France as early as 1924 with the foundation of the "Institut d'Urbanisme" at the Sorbonne in Paris (Albers, 1997).

Independent courses for "planning" – in addition to "design" – were created in particular in Germany and Austria towards the end of the 1960s and 1970s. In these countries this led to the splitting of disciplines into *urban design*, anchored in architecture and based on three-dimensional thinking, and the intrinsically interdisciplinary *urban and regional or spatial planning*, anchored in the social sciences and motivated by politics and government. Here urban design is regarded as merely one specialist area among many others – such as law, ecology, economy, sociology, etc. These understandings of the disciplines reveal differences in the cultures, mindsets and traditions of the various European countries. The concept of "urbanisme", coined in France in 1910 (Albers, 1997), spread out in the Napoleonically influenced European area (Nadin and Stead, 2008). Here understanding of the control of spatial development is characterised by planning understood as control of the construction of towns. With the concept of "town planning" in the United Kingdom or "Raumplanung" (spatial planning) in Germany and Austria, there prevails an understanding that design is separate from planning and is the domain of architects as "designers".

The societal and political framework for gendered perspectives in spatial development and planning

The state of knowledge and experience concerning gendered approaches in spatial development has been influenced by the history of the women's movements in modern times. Four "waves" can be distinguished within the women's movements.

The first wave emerged in the times of the French Revolution in 1789 (Gerhard, 2012). Since the eighteenth century, different women's and civil rights movements and debates had paved the way for the legislative implementation of women's suffrage and equity (e.g. 1866 Women's Suffrage Petition in the UK; Dohm, 1874; Gerhard, 2012). The first European country which implemented women's suffrage was Finland in 1906, followed by Austria in 1918, Germany and the Netherlands in 1919, the United Kingdom in 1928, Spain in 1931, France in 1944, Greece in 1952 and – rather late on the national level – Switzerland in 1971. Female students had access to universities in European countries since the beginning of the twentieth century, except for Switzerland where women had been allowed to register since the mid-nineteenth century. Legal status concerning gender equality is very important. Hence, according to Pascall and Lewis (2004), equal treatment across all five levels of autonomy and political intervention – voice, paid work, income, care work, time – has not yet been achieved in any of the European countries, as demonstrated, for example, by the "gender wage gap". Legal and political equal treatment has not been translated so far into the equal valuation of female and male working contexts and life situations (European Commission, 2014). Still, the equal treatment perspective sustains the implementation of equal rights and equal opportunities in European and national legislation.

Especially in the European context, the epoch-making work of Simone de Beauvoir's "The Second Sex" (Beauvoir, 1949) was the trigger for the second wave of the women's movement. It criticised patriarchy and focused on the differences between women and men as well as on differences in their lives and requirements (e.g. Döge, 2001; Wetterer, 2004; Damyanovic, 2007). The women's perspective criticised the one-dimensional orientation towards male norms. From this perspective women were always seen as "the other" so that they have to emancipate themselves from male norms in order to be individuals. Therefore, it was essential to highlight differences between women and men in terms of biological and physiological characteristics (see Chapter 1, Table 1.1). Certain male and female characteristics define the sexual order. In politics, this difference perspective was the basis of positive action in gender-equality policies (Booth and Bennett, 2002).

Debates about gender as a structural category and of changing gender relations influenced the third wave of the women's movements which started in the 1980s. Since then gender has also been understood as a societal category to overcome stereotypes based on biological and physiological characteristics and to emphasise the social construction of gender and gender relations (Chapter 1,

Table 1.1). The gender perspective has resulted in gender mainstreaming strategies and gender-sensitive policy analyses in research and practice in Europe. Debates on gender+ and intersectionality have broadened understanding of gender as a structural category (e.g. Knapp and Wetterer, 2007; Bustelo, 2017).

Gender as a process category finally refers to the construction and deconstruction of gender identities (see Chapter 1, Table 1.1). (De-)Constructivist perspectives, on the one hand, refer to the "Doing Gender" approach introduced by West and Zimmerman (1987) and, on the other hand, to "Gender Troubles" discussed by Butler (1991). The debate was also taken up in the European context (e.g. Becker and Kortendiek, 2004; Gildemeister, 2004; Kortendiek, Riegraf and Sabisch, 2019). Worldwide, there is growing evidence of a fourth wave of the women's movements: a turn towards solidarity politics and its reflection in self-organising groups, due to new possibilities of communication and networking as a consequence of digitalisation (Horelli, 2017; see also Chapter 6). The fourth wave has not yet been adopted by academic feminist planning research but becomes obvious and influential in the publications of bloggers and journalists, e.g. Laurie Penny (Penny, 2018) or Marisa Bate (Bate, 2018).

Important concepts and debates for gendered perspectives in spatial development and planning practice and research

Gendered perspectives in spatial development and planning are presented chronologically and with respect to predominant underlying concepts of gender: from the utopian feminists to women-centred approaches referring to gender as a difference category, to gender approaches referring to gender as a structural category, and finally to (un)doing gender approaches referring to gender as a process category.

The following section is written with a focus on German-speaking countries – for two reasons: on the one hand because of the considerable and also essential body of knowledge concerning women and gender in spatial development and planning that exists just here, in a central part of Europe, and on the other hand because these countries were particularly well represented within the International Working Group which is responsible for this book.

Utopian feminists and social movements

Until the end of the nineteenth century, debates focused on the social situation of women and men (in the sense of equal rights) in cities and on women's housing concepts to reconcile productive and reproductive work. They emerged at about the same time as the first European socialists, the so-called "social utopians", especially in France and the UK, began to generate ideal models of a good society and adequate new forms of community housing. Some among them, e.g. Charles Fourier, also adopted feminist positions. In the USA, the first feminist concepts concerning women's needs in residential building and town planning were created by the so-called "material feminists" (Hayden, 1981;

Greed, 1994; Vestbro and Horelli, 2012). They campaigned against women's isolation at home and their confinement to domestic life, seeing these as the basic cause of their unequal position in society. This resulted in far-reaching models of cooperative housekeeping by Melusina Fay Pierce or kitchenless houses by Marie Stevens Howland and, furthermore, in urban models that planned underground infrastructure by Alice Constance Austin. Cooperative housekeeping and kitchenless houses became the basis for the plan of a whole city, partly implemented in New Mexico (Topolobampo) in 1885 by M. Stevens Howland together with her male colleagues J.J. Deary and A.K. Owen (Hayden, 1977; Vestbro and Horelli, 2012). At the end of the nineteenth century there were similar – even though not as far-reaching – debates in European metropolises, implemented in the German-speaking world (e.g. Zurich, Vienna, Berlin) with the housing type "Einküchenhäuser" (one-kitchen buildings) (Pirhofer, 1978; Uhlig, 1981). Also, in the Scandinavian countries the concept of a "central kitchen" was discussed as a contribution to the *collectivisation* of domestic work (Vestbro, 1997). "In most nineteenth century communes 'women's work' remained sex stereotyped, but men and women benefited when cooking, cleaning and child care were collectivized" (Hayden, 1977: 25).

In the first few decades of the twentieth century, when women had attained the right to study, attempts at the legislative and social enhancement of the female situation emerged in the design work of early female architects (e.g. Roberts, 1991; Dörhöfer, 1994; Greed, 1994; Zibell, Damyanovic and Àlvarez, 2016). Lily Braun was one of the first architects in Germany who dealt with equalities through the reformation of kitchens (Braun, 1901; Schmidt-Waldherr, 1999). Still, the main planning issue was domestic work as a domain of women or as a reference point for emancipation. In this context, it is important to mention the Austrian architect Margarete Schütte-Lihotzky, one of the first internationally recognised female architects who worked (not only) on social housing projects in Austria, Germany, Russia and Turkey. Internationally she became famous with the so-called "Frankfurt Kitchen" in the 1920s (Lihotzky, 1927; Kuhn, 1998).[2] This concept was a turn to modernism and the *rationalisation* of domestic work, but the sex stereotype that women are responsible for reproductive work and men are the breadwinners remained.

Another topic in the field of housing, addressing specific sections of females such as elderly, single or pregnant women, single mothers or female students by internal housing layouts and urban design (Dörhöfer, 1990), also has its origin in the early contributions of female architects to architecture and housing. In a study about women's input to urban development, housing reform and architecture debates from 1870 to 1933, German sociologists examined the original contributions of women, at that time a quite lively topic of public discussion on, for example, housing projects for women and concepts about the collectivisation of domestic work. They provided evidence that such work became buried in oblivion or was absorbed by mainstream planning and housing reform debates without reference to its origin, even within the women's movement itself – especially after the decline of the first women's movement in the course of

Table 2.1 Concepts and debates of gendered approaches in spatial development and planning

	Feminist utopians and social movements	*Women-centred approach*	*Gender approach*	*(Un)Doing gender approach*
Timeline	End of nineteenth/ beginning of twentieth century	Since the 1970s	Since the 1990s	Since the 2000s
Analytic categories	–	Difference category (Sex)	Structural category (Gender)	Process category (Gender⁺)
Important debates in gendered approaches	Nineteenth century: utopian feminists, aiming at collectivisation of domestic work 1920s: the "New Woman" (in modern times), aiming at rationalisation of domestic work	Women-specific planning (*frauenspezifische Planung*) Women-friendly planning (*frauenfreundliche Planung*) Women-fair planning (*frauengerechte Planung*)	Gender-sensitive planning Gender-aware planning Gender (mainstreaming) planning	Performative planning Queer planning Self-organisation of groups facilitated by new digital communication
Results and products	Nineteenth century: radical overall concepts; kitchenless houses, co-housing projects Twentieth century: special housing for women, emphasis on social facilities	Criteria for women-fair (urban) planning, pilot projects in housing and building/urban design	Gender-sensitive planning as a cross-sectional and integrational issue	(Un)Doing gender as a psycho-social and socio-cultural trajectory permitting varying gender expressions and sexual orientations
Input to planning disciplines and theoretical framework	Theories and concepts, practice (bottom-up)	From theory to practice and pragmatism (women as experts for housing and neighbourhood)	Gender as an analytic tool and policy strategy (top-down)	Gender as a process (gender identity, queer, expression, orientation)
Transfer of gendered issues in urban planning/ spatial development since the twentieth century	Planning focusing on the political improvement of the social situation of all, housing concepts with emphasis on collective domestic work	Women-only places, women-fair housing, participation of women, needs and user-oriented planning, aware of underrepresented people	Planning with the awareness of differences; participation focusing on gender and diversity, planning for integrating societal production and reproduction spheres	Planning for gender variable and transgendered people, taking into account the variety of the non-heteronormative population, planning as a transformative process

World War I and, at the latest, during the Nazi regime (Terlinden, Grieser and Ross, 1999; Terlinden and Oertzen, 2006).

Women-centred approaches – gender as a difference category

Women-centred concepts in architecture and planning were introduced to a mainstream audience in the German-speaking countries for the first time at the end of the 1970s in a double issue of *Bauwelt*, the most widely read professional journal among planners and architects (*Bauwelt* 31/32, 1979). Women like Margret Kennedy, Paola Coppola Pignatelli, Cillie Rentmeister, Myra Warhaftig and Franziska Bollerey had the chance to present their specific perspectives on women in architecture. In the 1980s and 1990s the teaching of women's perspectives by female assistants as well as final theses by female students expanded the topics and research of feminist issues in spatial sciences (e.g. Hiort and Zibell, 1980, published by Zibell, 1983; Dörhöfer, Flecken and Zibell, 1998; Terlinden, 2010; Damyanovic and Fuchs, 2014). Also, professorships for women-centred architecture and planning were established (Zibell, 2019). The recent teaching and research project "On Stage – Women in Landscape Architecture and Planning" shows the history of this development and makes female architects and planning experts visible (Zibell, Damyanovic and Àlvarez, 2016).

Women-centred planning debates in the German-speaking countries have promoted three – also controversially discussed – approaches: women-specific planning (*frauenspezifische Planung*), women-friendly planning (*frauenfreundliche Planung*) and women-fair planning (*frauengerechte Planung*) (e.g. Kose and Lička, 1993; Becker, 1997; Dörhöfer and Terlinden, 1998; Grüger, 2000; Wotha, 2000). Important contributions to feminist planning were made by the periodical of FOPA (Feministische Organisation von Planerinnen und Architektinnen) called *FreiRäume* from 1983 to 2004.[3] Later, in 1993, the Austrian Landscape Planning Journal *Zoll+*[4] (Zeitschrift für Landschaftsplanung und Landschaftsarchitektur) introduced feminist approaches to landscape architecture and planning (Forum Landschaftsplanung, 1993, 2007). The beginning of the 1980s saw the first attempts to reflect on architecture by women in Switzerland (Schnitter, 1984). Later, feminist studies of the safety of women in public spaces were executed in various Swiss cities; these can be considered the first feminist analyses of cities, spaces and spatial appropriation (Karácsony and Zibell, 2018).

The debates about women's safety in public spaces were crucial for efforts to create women-friendly cities (Dame and Grant, 2001).[5] When enhancing safety in public spaces through physical factors (e.g. street lighting), social factors (e.g. the presence of different user groups), functional factors (e.g. mixed-use neighbourhoods) and building factors (following the principle "eyes on the street" introduced by Jane Jacobs as early as 1961), subjective factors are important as well (e.g. Siemonsen and Zauke, 1991; Miko, Neureiter and Stadler-Vida, 2012). A big debate dealt with intimidating public spaces, so-called "Angsträume" (Sailer, 2004; Becker, 2008) because this term suggests that the space has agency for felt uncertainty, not men. For urban public spaces, girl-centred concepts (girls' parks) were professionally developed and implemented especially in the City of Vienna (Studer, 2002; Tilia, 2007), one of the most densely built cities in Europe.

An important topic continues to be the broad field of housing – for the use of women as well as conceived and built by women. Thus, the women-centred approach promoted women's issues such as support for autonomous women, elderly women, women with lower incomes, single mothers or collective women's projects (Rebe, 2002; Becker, 2009). Practical knowledge for women-specific or women-adequate housing was gained by projects such as Frauen-WerkStadt I in Vienna, with about 360 housing units the biggest in Europe (Zibell and Schröder, 2007; Kail and Irschik, 2013; Ullmann, 2013; Riss, 2017). Other projects which contributed to enhancing practical knowledge in this context were Brahmshof in Zurich (Zibell, 1993; Schröder and Zibell, 2004), smaller projects in Linz, Salzburg and Graz (for Austria) (Riss, 2017) and (for Germany) Freiburg i.Br., Bergkamen (Schröder and Zibell, 2004) and Beginenhöfe in Berlin (Diesenreiter, 2015).

At the same time, checklists with criteria for designing and evaluating housing projects and, even more important, the quality of urban life in general, supported the everyday needs of different genders, ages and social backgrounds, benefiting all the urban population. The checklist criteria facilitate housework and family tasks, promote good neighbourly contacts, foster safe environments and include provisions for various accommodation for different stages of life as well as an adequate range of semi-private and semi-public open spaces (Altenstraßer, Hauch and Kepplinger, 2007; Zibell and Schröder, 2007; Vestbro and Horelli, 2012; Kail and Irschik, 2013; Tummers, 2013).

The findings of women-centred approaches have contributed to theoretical debates on women's issues in planning and to practical criteria and guidelines not only in the fields of housing, public spaces and safety (e.g. Siemonsen and Zauke, 1991; Frauenlobby Städtebau, 1993; Kramer, 2002) as well as public transport and safety (e.g. Bauhardt, 1995; Knoll, 2008), but also in the context of general principles such as sustainable urban and regional development (e.g. SRL FG Frauen, 1999; Grüger, 2000) and the European city as a model (Zibell, 2013).

An important contribution initiated by the German federal institute BBR (Bundesamt für Bauwesen und Raumordnung)[6] was made by the project "Frauengerechte Stadtplanung" (women-fair urban planning), which marks a turning point within the women-centred debates by introducing the term "gender-sensitive" planning in the German context (Baumgart and Seggern, 1996). A round table in Bonn initiated by the active regional FOPA Hamburg network in collaboration with the SRL Fachgruppe Frauen (women expert group) prepared and published women's strategies for sustainable urban and regional development for the world conference "Urban 21", which took place in Berlin 2000 (SRL FG Frauen, 1999). Both attempts – woman-fair urban planning as well as women's strategies for sustainable spatial development – were not further pursued in the face of the up-and-coming gender-mainstreaming strategy, neither have they been generally adopted within the mainstream of spatial development and planning practice or theory to this day.

The women-centred debates also received criticism in research and practice as they were seen as white and middle-class oriented women-only planning, in particular ignoring disabled, black and minority ethnic or poor people. It was considered that they contributed to maintaining the status quo of gender relations instead of changing them in the direction of gender equality and optional ways of life for all genders (Kose and Lička, 1993; Becker, 1997; Dörhöfer and Terlinden, 1998; Wetterer, 2004).

Gender approaches – gender as a structural category

Feminist planning debates were put thoroughly to the test and subsequently renewed by gender approaches perceiving gender as a structural category and by the political strategy of gender mainstreaming (Chapter 1). Thus, the gender debate no longer focused on highlighting women's ideas and concerns. Instead, the focus turned to structural imbalances between genders becoming apparent in the city, village or region as a whole (Terlinden, 1990; Bock, Heeg and Rodenstein, 1997), to access to spaces and resources, and to giving voice to underrepresented people in equitable planning and participation processes (Damyanovic and Zibell, 2013). Debates on unequal gender power relations have been selectively discussed (Terlinden, 1980, 2006; Kail and Kleedorfer, 1991; Scheller, 1995; Kuhlmann, 2002) but have to date been inadequately implemented in research and insufficiently embraced by mainstream spatial and planning theories (Zibell, 2019).

In the 1990s, when gender became a cross-sectional issue in politics and planning debates in Europe (e.g. Horelli, Booth and Gilroy, 2000; RTPI, 2007) and in the Global South (Moser and Levy, 1986), several planning academics from Europe and the USA influenced the gender-planning discourse (Damyanovic and Zibell, 2013). For example, *Readings in Planning Theory* (Campbell and Fainstein, 1996) mentions that gender dimensions transform topics in urban planning, such as safety aspects in cities or differing use of urban spaces. Feminist experts in planning research called for the "non-sexist" (Hayden, 1981; Rodenstein, 1994) or "gendered" city (Roberts, 1991; Greed, 1994; Fainstein and Servon, 2005; Fenster, 2005; Jarvis, Kantor and Cloke, 2009). The Research Group for the New Everyday life contributed to the debate on the reconciliation of public and private life through domestic work by women and men, emphasising the context from vicinity to neighbourhood also from a glocal perspective (Horelli, Booth and Gilroy, 2000; Vestbro and Horelli, 2012).

Including the European context, several case studies in urban planning and spatial development resulted in practical manuals for planning and design focusing on the implementation of gender concerns and perspectives in planning instruments and processes in the fields of urbanism, architecture, open spaces and mobility. They also led to the compilation of criteria lists for the analysis (and evaluation) of urban and rural spaces as landscapes for everyday life considering gender+.[7]

In Germany the applied research project "Gender Mainstreaming im Städtebau" (Gender mainstreaming in urban planning), funded within the framework of the national programme "Experimenteller Wohnungs- und Städtebau – ExWoSt" (Experimental housing and urban development) (BBR, 2006), attracted attention. This research carried out from 2003 to 2005 addressed the implementation of gender mainstreaming in local authority planning practice. Another research project "Gender Mainstreaming im Wohnungswesen" (Gender mainstreaming in housing), also within the framework of ExWoSt and carried out from 2004 to 2005, was discontinued at an early stage and remained unpublished because a federal reform of housing responsibilities resulted in a shift of competences from state to federal level.

A hot topic in the German-speaking world are the results of a study entitled "30 Jahre Gender in der Stadt- und Regionalentwicklung" (30 years of gender in urban and regional development) (Bauer and Frölich v. Bodelschwingh, 2017), which was commissioned by Deutsches Institut für Urbanistik Difu (German Institute of Urban Affairs) and investigated municipal and regional planning practices in Berlin, Munich, Vienna and the Ruhr area. An outcome of the study is that gender can be described as a success story but only for the above-mentioned cases. The study is not representative because it only reviewed the "spearhead" of gender-sensitive planning implementation and experience within the German-speaking countries. Despite the acknowledgement that gender – whether implicitly or explicitly – could become more or less integrated in the routine work of planning, the difficult question is what will happen in the future when the personalities who supported and carried forward these processes retire. Until it is fully institutionalised, the implementation of gender mainstreaming in planning stands and falls with the commitment of personalities in leading positions in planning administrations. It remains to be seen whether the relatively positive assessment of this study will prove sustainable in the more distant future.

Academic debates also reflect the opportunities and risks of this strategy for the implementation of gender equality issues in various space-related sciences like landscape planning or regional development (e.g. Aufhauser *et al.*, 2003; Oedl-Wieser, 2006; Damyanovic, 2007, 2013; Protze, 2009), traffic-planning (Knoll, 2008), geography (e.g. Wotha, 2000; Wastl-Walter, 2010; Bauriedl, Schier and Strüver, 2010; Wankiewicz, 2013) and architecture, urbanism and spatial planning (e.g. Reeves, 2002; Greed, 2005; Schröder, 2009; Wotha, 2013; Horelli, 2017). Gendered approaches are, to a certain extent, included in the teaching of planning at universities in the German-speaking countries (e.g. Terlinden, 2010; Damyanovic and Fuchs, 2014; Zibell, Damyanovic and Àlvarez, 2016; Huning and Mölders, 2017; Zibell, 2019). The international teaching network Athena has published the handbook *Teaching Gender, Diversity and Urban Space* (Lada, 2009).

There has also been some criticism of the top-down policy of gender mainstreaming (Greed, 2005; Sainsbury and Bergqvist, 2009; Bacchi, 2011). One of the most commonly expressed opinions is that the policy strategy of gender mainstreaming is not transformative (Roberts, 2013). The gendered approach is to be applied more as an integrationist approach which entails the equal inclusion of women and men (*equality perspective*) and the participation of every gender in planning projects and processes (Bradshaw, 2015).

(Un)Doing gender approaches – gender as a process category

The gendered approach as a process, which focuses both on the integration of gender at all stages of planning as well as on analysing and addressing power relations, has rarely been part of the mainstream in planning sciences and planning practice (Roberts, 2013). Especially in cases of advocate planning where professionals act as agents for underrepresented people's interests, it seems very difficult to deconstruct the binary of female and male identities. (Un)Doing gender is a psycho-social and socio-cultural trajectory which contributes to constructing and deconstructing female and male identities (Chapter 1). Recently, there have been attempts to (de)construct gender identities, expression and orientation in planning and participatory processes (Damyanovic, 2007; Oswin, 2008; Strüver, 2018). The scientific and – even more so – the practical mainstream planning debates find themselves at the very beginning of this discourse.

As mentioned before, the new possibilities of communication and networking resulting from digitalisation may have an important influence on the (de-) construction of gender identities (Horelli, 2017; see Chapters 1 and 6). Queer theories (Oswin, 2008; Wastl-Walter, 2010; Hofmeister, Katz and Mölders, 2013; see Chapter 1) criticise the standardisation of gender identities and visualise people, actions and things deviating from the norm of dual sexuality (Mölders, Kanning and Zibell, 2014; Gildemeister, 2019). The book *Planning and LGBTQ Communities*, edited by Petra L. Doan (2015), discusses and examines diverse approaches to queer planning, the variety of non-heteronormative populations and the different ways that queer people live. The authors refer to the discussion of the need of LGBT (Lesbian, Gay, Bisexual and Transgender) communities for safe urban spaces. But there are still many debates to come in the practice and research of spatial planning and development in order to incorporate gender as a process category.

Gendered approaches in the context of spatial planning theory

From the authors' perspective, the integration of gendered approaches in spatial planning theories and academic debates is completely insufficient. This claim is supported by reference to two popular books on planning theory in Europe (see Damyanovic and Zibell, 2013): *Planning Theory* by Philip Allmendinger (2002), Professor of Land Economy, and *Planning in Crisis?* (*Planung in der Krise?*) by Walter Schönwandt (2002, 2008), Emeritus Professor for Planning Theory. Both authors describe the current landscape of planning theory by modelling a typology of theoretical approaches (i.e. rational, Marxist, neo-liberal, communicative, collaborative, equity and advocate planning). A comparison shows that both discuss similar planning issues in comparable contexts. Gender and women's issues, however, remain disregarded, although they have already found their way into academic thinking and feminist planning (see the paragraph on gender approaches above). In their analyses, Philip Allmendinger and Walter Schönwandt identify different planning approaches as a basis for different theories in planning, but women-centred or gendered approaches are completely absent (Damyanovic and Zibell, 2013).

Clearly, gendered approaches have not so far found their place within the planning mainstream. This volume makes another effort to bridge the gap between gender and mainstream planning approaches by trying to reconcile gender theories and methodologies concerning spatial development. But it remains a field requiring good political will – gender relations cannot be changed by spatial development or planning alone. The book contributes to values and planning principles which support a gender-equal planning culture. Nevertheless, it is obvious that modified perspectives on space and spatial development and openness to "other" or hybrid realities broaden the possibilities for a diversity of options/optionality and freedom of deployment for every citizen, user and resident.

Leaving the historical perspective, the following section is dedicated to present-day European planning systems and their connection to gender issues.

Selected European perspectives: planning systems and gender issues in nine European countries

Stephanie Tuggener and Barbara Zibell

Planning systems framing spatial development

As stated in Chapter 1, we understand spatial development "as a field of action and governance among politics, planning, economics and people". A country's planning system and its self-conception are central frameworks and decisive factors for (a society's practices in) spatial development (see Albers, 1997) – and vice versa. A planning system with laws and policies, institutions and instruments is the result of processes of political negotiation in a specific cultural context. Societal norms and values as well as structures of power manifest themselves in a planning system. It stands for the formal, institutionalised powers that steer spatial development. Besides, there are a variety of other forces and players such as informal bottom-up movements, societal organisations or economic forces influencing spatial development.

This chapter focuses on national planning systems with their corresponding laws, policies and instruments since their influence on spatial development is explicit and powerful. The goal of this section is to shed light on the questions of whether, where and how gender plays a role in different planning contexts and which roles for gender can be identified in them. There are two core research questions: Are the planning systems gender-sensitive? Is there a gender impact (institutionalised or not) on spatial development? The authors compare nine European countries' planning systems with regard to the role and the significance of gender. For a better overview, the information is presented in factsheets (see Tables 2.2 to 2.10).

Case-study countries

The authors studied the planning systems of Austria, England, Finland, France, Germany, Greece, the Netherlands, Spain and Switzerland. This selection reflects the origin of the members of the international working group responsible for this book. Except for Switzerland, all analysed countries are EU members. This shared context brings about common basic conditions, like for example spatial policies or subsidies with spatial impacts such as EU structural funds and the European Spatial Development Perspectives (ESDP[8]). Moreover, the strategy of gender mainstreaming has been a binding principle for all EU members since 1997 (Treaty of Amsterdam). In Switzerland, which adopted the Treaty of Amsterdam, a lot of activities to enhance gender equality are also undertaken.

National factsheets – structure and content

In the following section, the structure of the national factsheets is briefly described. The authors are fully aware that the tables are a simplification and that the categories chosen do not fully reflect the complexity of any of the countries' planning systems.[9] Nevertheless, the authors are convinced that this attempt to structure and compare the countries is helpful to enhance understanding of the current situation of gender in spatial planning.

In order to contextualise the various national planning systems and to characterise the underlying political systems, some basic facts are presented first. Then the planning system as the central framework and decisive influencing factor is analysed in more detail. This characterisation allows an analysis of whether gender aspects appear on a formal planning level (national, federal, regional, local) or are part of a planning law, policy or formal planning instrument.

Each country's most important gender impacts are stated separately. The authors understand this impact broadly, ranging from laws, policies and regulations to projects and activities relevant to spatial development that take gender explicitly into account or aim to strengthen gender equality. Correspondingly, the pursued objectives of these "gender impacts" can be very diverse. The authors subsume gender-sensitive laws, pilot projects, manuals, scientific case studies and activist movements in the category "Gender impact on spatial planning". Finally, they describe if and what kinds of gender-related networks exist in the different countries.

Since gender-related issues frequently appear only implicitly or are integrated in substitute concepts such as social sustainability or options for public participation, these aspects are studied as well. A society that takes the concept of sustainable development really seriously would weigh the social, ecological and economic needs evenly. Consequently, it would work on a transformation to overcome hegemonies and to become a fair, cooperative and in this spirit also gender-sensitive society. This applies similarly to the possibilities for participation. A society that takes every citizen seriously and regards everybody as being of equal worth would establish fair and cooperative processes that are also gender-sensitive.

The order of the factsheets is oriented to the classification of Sapir (2006; cited in Nadin and Stead, 2008), concerning typologies of social order in Europe: first the Nordic countries (FI, NL), then the Anglo-Saxon (UK), the continental (AT, DE, FR) and the Mediterranean (ES, GR); Switzerland as a non-EU member is placed at the end.

National perspectives of nine European countries: planning systems and gender impact

Table 2.2 National perspective of Finland

Political framework

- centralist state, unitary system: power resides with the national government, some responsibilities may be delegated to specific departments or local government units
- bi-polar structure (national/local); weak governance structures at the regional level

Planning system

- comprehensive/integrated type: normative, multi-level system, focusing on land-use planning
- strong sector orientation dividing the various parts of spatial development policy into separate policy fields
- planning is rooted in the field of architecture and separated from environmental and regional development policies

Levels of spatial planning

Laws/policies on spatial planning	*Planning instruments*	*Organisations and main actors in spatial planning*
National level		
• Town Planning Act 1932 (*Laki taajaväkisistä maalaisyhdyskunnista*): establishment of zoning system • Building Act 1958 (*Rakennuslaki*): establishment of planning hierarchy • Land-Use and Building Act 1999 (*Maankäyttö-ja rakennuslaki*): basic building right, right to participate in planning processes, principle of sustainability • Regional Development Act 2009 (*Laki alueiden kehittämisestä*): defines goals of regional development and tasks of responsible authorities	• National Land-Use Guidelines (*Valtakunnalliset alueidenkäyttötavoiteet*): advisory, guidance policy on land-use issues • National Regional Development Targets (*Valtakunnalliset alueiden kehittämistavoitteet*): specify goals of the Regional Development Act	• government • national planning agency
Federal state level		
–	–	–
Regional level		
–	• Regional Plan (*Maakuntakaava, maakuntasuunnitelma*): binding, directs land-use planning at the local level, adjusts national, regional and local aims	• 18 regional councils (*Maakunnan liitto*)

Local level		
–	• Joint Master Plan (*Kuntien yhteinen yleiskaava*): by two or more municipalities, optional • Local Master Plan (*Yleiskaava*): determines overall land-use framework, optional • Detailed Plan (*Asemakaava*): defines binding building rights, dimensions and functions, mandatory	• municipal councils (*Kunnanvaltuusto/ kaupunginvaltuusto*)

Gender impact on spatial planning

- The Act on Equality between Women and Men (1986) every authority to enhance gender equality in their work; the act is poorly known and badly implemented
- Government Action Plan for Gender Equality 2016–2019 does not mention spatial development; but: "all ministries will work on gender equality and investigate key challenges to equality in their policy areas" (Ministry of Social Affairs and Health, 2017: 19)
- Finnish Land-Use and Building Act does not mention gender issues; however, there is a reference to the needs of different user groups (young, elderly, etc.)
- there are official equality institutions at governmental level; due to the lack of actors with gender expertise, equality competences are not represented in the decision-making bodies at regional level
- the development work for gender mainstreaming is coordinated by the Association of Finnish Local and Regional Authorities (*Kuntaliitto*)

Specific perspectives: "gender blindness"

- Finland is currently a gender-neutral country, the so-called "gender contract" implies that men and women are already equal and therefore no further measures are required; the "gender contract" (Hirdman, 2001) refers to normative cultural relations, processes and action spaces of women and men in different spheres/levels
- thinking that equality between women and men has already been reached leads to the assumption that a gender perspective is redundant
- because of "gender-blindness" gender has no influence on spatial development

Gender-related networks

- none in the field of spatial planning or development
- women's Facebook group "sewing-club built environment"

Table 2.3 National perspective of the Netherlands

Political framework		
• unitary, top-down, with provincial structure • parliamentary democracy and constitutional monarchy		
Planning system		
• three levels of government (provincial, regional, local) • regional "*Waterschappen*" (water authorities) • the "polder model" of negotiation characterises the planning process: area development through co-creation • several "regional boards" have been created (bottom-up) as self-appointed planning bodies through municipal initiatives; they have no legal basis but operate effectively • structure visions and development plans for cities and metropolitan regions • national strategies for subsoil planning; flooding risk; infrastructure		
Levels of spatial planning		
Laws/policies on spatial planning	*Planning instruments*	*Organisations and main actors in spatial planning*
National level		
• Spatial Planning Act (*Wet ruimtelijke ordening Wro*): states how spatial plans are developed and who (province, municipality) is responsible • Spatial Planning Decree 2017 (*Besluit ruimtelijke ordening Bro*): further elaborates *Wro* and sets general rules for spatial planning; will be replaced in 2019 by the Environment Act (*Omgevingswet*)	• Structural Vision (*Structuurvisie*) • National structure visions (*Vierde nota ruimtelijke ordening VINO* 1988, *Vierde nota ruimtelijke ordning extra VINEX* 1991) • Structural Vision Infrastructure and Space 2040 (*Structuurvisie Infrastructuur & Ruimte SVIR*), with investment programme infrastructure, space and transport (MIRT) • Guideline for Sustainable Urbanisation (*Handreiking duurzame stedelijke ontwikkelingen*)	• government • main tasks: supervision of provinces and municipalities
Federal state level (Provincies)		
• *Wro/Bro* • granted considerable autonomy	• Structural Vision (*Structuurvisie*) • Area Development Plans (*Gebieds-ontwikkelingsplannen*) • Urban Regional Plans (*Streekplannen, Structuurplannen*)	• provincial authorities; councils (*Provinciale Staten*) and executives (*Gedeputeerde Staten*) • main tasks: town and country planning, public transport, environmental water management, welfare, economic and agricultural matters, supervising municipalities

Regional level		
• granted considerable autonomy	• Area Development Plans (*Gebiedsontwikkelingsplannen*): developed locally and in regions • Urban Regional Plans (*Streekplannen, Structuurplannen*)	• several regional boards without legal basis
Local level (Gemeenten)		
• *Wro/Bro*	• Structural Vision (*Structuurvisie*) • Area Development Plans (*Gebiedsontwikkelingsplannen*) • Land-Use Plans (*Structuurvisie, Structuurplannen*) • Zoning Plan (*Bestemmingsplan, incl. Stadvernieuwingsplan*) • Building Licences (*Omgevingsvergunning*)	• municipal authorities (council and executives) • deal with matters that affect residents directly and exclusively, e.g. provide housing, construct local streets

Gender impact on spatial planning

- no implementation of gender in spatial legislation or regulation beyond the general legal framework (constitution) which assures equal rights and the Equal Treatment Law of 1994
- historically, influence of gender in spatial development appears mostly bottom-up, e.g. criteria for safety in public space (now generally accepted as an issue); consulting the expertise of user-groups (united in *Adviescommissie voor de woningbouw*) for the design of new houses and neighbourhoods
- currently, most activities to enhance gender equality focus on academia or entrepreneurship
- awareness of gender mainstreaming in society is very low and in spatial development practices even lower; little has remained from the initiatives of the 1990s (despite good results of the European Social Fund's daily routines scheme)
- EU obligations are implemented top-down (National Government), all other activities are by NGOs and feminist/women's networks and initiatives
- gender studies persists in some universities, mainly in social sciences (Nijmegen, Utrecht, Amsterdam)

Specific perspectives

- politics focuses on labour market issues and some superficial policies for "diversity"
- emancipation, gender equality, diversity are used interchangeably and with different concepts

Gender-related networks

- *Bouwnetwerk*: network of female professionals in the building industry
- Delft Women in Science (DEWIS)

Table 2.4 National perspective of England

Political framework

- England is part of the United Kingdom which also includes Wales, Scotland and Northern Ireland; the Westminster Parliament has sovereign (ultimate) power over the UK as a whole
- spatial planning in different parts of the UK is organised differently; because of the origin of the working group members who are responsible for this book the focus is on England

Planning system

- a system which cascades down from central government, from general principles and policies to detail at local municipal level
- planning law is set by precedent as well as by published policy and legislation
- policies at local level have to be in accordance with those at national level
- planning is a quasi-legal system and involves negotiation for major applications at local level

Levels of spatial planning

Laws/policies on spatial planning	*Planning instruments*	*Organisations and main actors in spatial planning*
National level		
• Planning and Compulsory Purchase Act as amended • Town and Country Planning (England) • Regulation 2012 as amended • Localism Act 2011 • The Neighbourhood Planning (General) • Section 106 Town and Country Planning Act 1990 • Planning (Listed Buildings and Conservation Areas) Act 1990	• National Planning Policy Framework (NPPF) (Revised July 2018)	• Ministry of Housing and Local Government • Planning Inspectorate (PIN) • advisory bodies (e.g. English Heritage)
Federal state level		
–	–	–

Regional level		
–	• London Plan (Greater London only) includes Opportunity Area Frameworks and Supplementary Planning Documents • outside of London authorities have a "duty to cooperate" across boundaries	• Mayor of London and Greater London Assembly
Local level		
–	• Local Development Schemes or Frameworks: incorporating Local Plan, Statement of Community Involvement, Site Allocations, Environmental Impact Assessment, Equalities Impact Assessment, Supplementary Planning Documents, Conservation Area Plans, Area Action Plans, Neighbourhood Plans, Article 4 directions	• outside of London: county councils, district councils, town councils, parish councils • the main planning powers reside at district level • outside of London: also unitary authorities • London: London Boroughs, Westminster City Council and Corporation of City of London, designated public/private development corporations in Opportunity Areas

Gender impact on spatial planning

- all authorities are required to comply with the duties set out in the Equalities Act of 2011, "protected characteristics" are race, disability, gender, age, sexual orientation, gender reassignment and religion or belief
- plans and major planning applications are assessed at development/application stage, i.e. prior to implementation
- in the 1980s a few planning authorities set up Women's Committees, but these have been disbanded as equalities legislation has reached the statute book
- specialist interest groups such as the Women's Design Service have not been able to survive because of lack of funding
- recently the Women in Planning Network and Planning Out have formed to represent women planners and LGB (lesbian, gay, bi) planners
- Burgess's study of the impacts of the Gender Equality Duty which preceded the Equalities Act demonstrated little understanding of implementation and evaluation (Burgess, 2008)

Gender-related networks

- Women in Planning
- Urbanistas
- Equilibrium Network

Table 2.5 National perspective of Austria

Political framework

- federal republic
- nine federal states (*Länder*) have considerable autonomy from national government

Planning system

- decentralised planning system
- municipalities are the main planning authorities
- every federal state has its own legal planning framework (nine different planning and building laws) and a federal system of housing subsidies
- emerging policies for cross-border planning strategies in city regions based on voluntary cooperation plans and EU-financed development strategies
- each municipality/city has its own development concept and land-use plan

Levels of spatial planning

Laws/policies on spatial planning	*Planning instruments*	*Organisations and main actors in spatial planning*
National level		
• no national planning legislation • planning competences and some sectoral policies in the fields of superordinate street system, railways, navigation, forestry or energy infrastructure	• Austrian Spatial Development Concept 2011 (*Österreichisches Raumentwicklungskonzept* ÖREK): guiding principles for spatial development strategies, governance and cooperation	• national government: restricted responsibilities with strong focus on infrastructure, finance • Austrian Conference on Spatial Planning (*Österreichische Raumordnungskonferenz* ÖROK): coordination board
Federal state level (Länder)		
• nine different planning acts (*Raumordnungsgesetze*) • nine different building codes (*Bauordnungen*) • additionally legislation for nature protection, housing, culture, youth, care facilities, health	• State Development Concept (*Landesraumordnungsprogramm*): spatial strategy • sectoral programmes, e.g. traffic, housing subsidy system and location development strategies	• state government with sectoral competences, e.g. in spatial/traffic • environmental planning, energy, climate change strategies, agriculture, nature protection • districts (some competences such as water, forest)
Regional level		
• Regional Planning Act (*Raumordnungsgesetz*) • regional spatial strategies, developed top-down by the federal state or bottom-up by municipal cooperation • EU-driven policies to develop regional strategies and distribute EU-funds	• regional or sectoral development concepts for parts of federal states (*Regionale Entwicklungsprogramme*) • 77 Local Action Groups (LAG) with local development concepts (*Lokale Entwicklungsstrategien LES*)	• state government: responsible for regional development concepts (in cooperation with municipalities) • Local Action Groups: representatives of local public institutions, private associations, companies, persons, regional development agencies/ associations for regional development projects (quota of 40 per cent women)

Local level		
• Spatial Development Concept and Land-Use Plan (*Raumentwicklungskonzept und Flächenwidmungsplan*): competence of the municipality to decide on local development	• City or Local Development Concept (*Örtliches/Räumliches Entwicklungskonzept*): strategic planning instrument • Land-Use Plan (*Flächenwidmungsplan*): legally binding land-use plan • Building Control Plans (*Bebauungspläne*): legally binding, not obligatory • Building Permits (*Baugenehmigungen*)	• mayor (1st instance) • municipal council (2nd instance)

Gender impact on spatial planning

- framework for the implementation of gender equality and equal opportunities: Treaty of Amsterdam, national equal opportunities and antidiscrimination laws
- ministerial council decision in 2000 on the implementation of gender mainstreaming in administration, Article 7 (2), Federal Constitutional Law
- anchoring of gender budgeting in 2009, Article 13 (3), Federal Constitutional Law
- decisions on gender mainstreaming and gender budgeting on federal state level between 2000 and 2009 (all nine federal states)
- Austrian Spatial Development Concept does not mention gender mainstreaming; chapter on "Social Diversity and Solidarity" mentions gender+ related topics: e.g. diverse needs of women and men; equal access; good design/planning for all
- due to the decentralised planning system, there are diverse approaches to gender in spatial development in each federal state and municipality (mostly pilot projects, no structural implementation, sometimes strong resistance against "gender issues")
- only 4 of the 12 biggest Austrian cities and state capitals mention gender mainstreaming explicitly in their spatial development concepts

Gender-related networks

- *Interministerielle Arbeitsgruppe für Gender Mainstreaming/Budgeting (IMAG GMB)*, 2000: inter-ministerial working group on gender mainstreaming/budgeting
- *Verein zur Vernetzung und Stärkung von weiblicher Expertise in den Bereichen Mobilität, Umwelt, Energie und öffentlicher Raum (WIMEN)*: association for networking and strengthening women's expertise in the areas of mobility, environment, energy and public space
- Network of Austrian Counselling Centres for Women and Girls: umbrella organisation of 55 centres in all nine federal states
- *Ziviltechnikerinnen Mitte (ZIMT)*: interdisciplinary working group within the professional association of civil engineers in Salzburg and Upper Austria
- *Genderplattform*: Austrian Platform of Research and Studies on Women and Gender

Table 2.6 National perspective of Germany

Political framework

- federal republic; main principles: democracy, rule of law, social state
- 16 federal states (*Bundesländer*) of which 13 "area" states and three "city" states with considerable autonomy from central government

Planning system

- organised as a decentralised, multi-level system within the framework of the German federal structure, consisting of supra-sectoral and sectoral planning
- hierarchically structured, influenced by three principles: subsidiarity (decision on the lowest level possible); constitutionally guaranteed municipal self-government; mutual feedback principle

Levels of spatial planning

Laws/policies on spatial planning	*Planning instruments*	*Organisations and main actors in spatial planning*
National level		
• Federal Spatial Planning Act (*Raumordnungsgesetz ROG*): defines supra-local spatial planning, instruments and cooperation • Federal Building Code (*Baugesetzbuch BauGB*): defines local spatial planning, instruments and processes	• Spatial Planning Policy Guidelines (*Raumordnungspolitischer Orientierungsrahmen*): federal action framework • Spatial Planning Report (*Raumordnungsbericht ROB*) • Sectoral plans, e.g. Federal Transport Infrastructure Plan	• national government • Conference of the Ministers of Land-Use Planning (*Ministerkonferenz für Raumordnung MKRO*) • Federal Institute for Research on Building, Urban Affairs and Spatial Development (*Bundesinstitut für Bau-, Stadt- und Raumforschung BBSR*)
Federal state level (Bundesländer)		
• 16 different State Spatial Planning Acts (*Landesplanungsgesetze LplG*): govern spatial planning within states • 16 different State Building Regulations (*Landesbauordnungen LBO*): provisions relating to construction on site, building permission procedures and the organisation of building authorities	• State Spatial Planning or Development Plan/Programme (*Landesraumordnungsprogramme LROP/Landesentwicklungspläne/-programme LEP*): outline the desired spatial and structural development for the territory of the state	• federal state government
Regional level		
–	• Regional Spatial Structure Programme (*Regionales Raumordnungsprogramm RROP*): outlines the spatial and structural development of regional planning areas	• partly administrative districts (*Regierungsbezirke, Regierungspräsidien, Bezirksregierungen*) • regional planning is organised in many different ways on the level of administrative districts (*Regierungsbezirke, Landkreise, Planungsregionen*)

Local level		
–	• Preparatory Land-Use Plan (*Flächennutzungsplan FNP*): sets out the municipality's future land use; preliminary representations on the use of plots • Binding Land-Use Plans (*Bebauungspläne B-Plan*): legally binding rules for the development and organisation of sections of the municipal territory	–

Gender impact on spatial planning

- progressive good practice in spatial planning only in a few cities such as Berlin, Munich, Ruhr area (Bauer and Frölich v. Bodelschwingh, 2017)
- the need for gender in spatial development is not regularly recognised
- explicitly implemented in the Federal Building Code (*BauGB*) (equal opportunities for women and men as reason of consideration in land-use planning)
- broad range of guidelines/criteria catalogues from the 1990s, supposedly universally applicable for planners in various planning sectors, mostly with the topic "Women and Urban Planning" (see Zibell and Schröder, 2007)

Specific perspectives: selected manuals

- Gender in Mainstreaming Urban Development (2011): Berlin on the path towards becoming a metropolis worth living in for women and men
- Gender Compass Planning (2009): Guideline on gender planning with respect to spatial quality and legal framework

Specific perspectives: perception of gender today

- gender equality is accepted as a desirable goal among many social groups and media
- there is a broad array of opinions on how far gender equality has already been reached
- in statistics, gender inequalities are clearly represented (gender pay gap, shares in higher career strata and governing bodies, choice of profession, shares in part-time work, time spent for housework and care work, etc.)
- currently a new wave of an anti-feminist movement, abusing gender research and threatening gender researchers personally, can be observed

Gender-related networks

Formal institutions (examples)

- expert women's advisory council of the Senate Department for Urban Development and Housing in Berlin
- Commission Women in the City at the German Association of Cities

Informal institutions (examples)

- FOPA Feminist Organisation of Women Planners and Architects
- Women's Network Ruhr Area
- Female Planning Experts (*PlanungsFachFrauen*), Region Hannover

Table 2.7 National perspective of France

Political framework

- centralist state
- unitary system: power resides with national government, some responsibilities may be delegated to specific departments/local government units

Planning system

- planning system is very close to the "regional economic ideal type" (EU Compendium; European Commission, 1997: 36)
- planning authority lies with the central state, national laws are central
- regions have little regulatory power
- there are no strategic national plans, regions are entitled to undertake plans
- system is characterised by broad aims (from regulating the use of space to social and economic aims such as reducing regional disparities)

Levels of spatial planning

Laws/policies on spatial planning	*Planning instruments*	*Organisations and main actors in spatial planning*
National level		
• Law on Spatial Planning and Sustainable Development, 1995/1999 (*Loi d'orientation pour l'aménagement et le développement durable du territoire* LOADDT; *"Voynet Act"*) • Act for Urban Solidarity and Renewal 2000 (*Loi relative à la solidarité et au renouvellement urbains*) • Bill for National Environmental Engagement 2010 (*Loi portant engagement national pour l'environnement, "Grenelle II Act"*) • Law on Access to Housing and Regeneration 2014 (*Loi pour l'accès au logement et à un urbanisme rénové* ALUR) • Law on the Evolution of Housing, Planning and Digital Technology 2018 (*Loi evolution du logement, de l'aménagement et du numérique* ELAN)	• Shared Services Scheme (*Schéma de Services Collectifs SSC*): different sectoral guidelines, which are not necessarily coordinated; there is no national spatial planning perspective	• government • Interministerial Agency for Spatial Planning and Competitiveness, DIACT (*Délégation interministérielle à l'aménagement du territoire et à l'attractivité régionale* DATAR)
Federal state level		
–	–	–

Regional level (18 "régions" of which 13 are metropolitan regions, subdivided into "départements")		
–	• Regional Scheme of Planning, Sustainable Development and Territorial Equality (*Schéma régional d'aménagement, de développement durable et d'égalité des territoires SRADDET*)	–
Local level (Communes or "intercommunalités" (metropolitan regions, city regions, agglomerations, single communities))		
–	• Groups of municipalities (*Schéma de cohérence territoriale SCoT*): sustainable spatial development perspective, strategic plan, top of hierarchy for local documents • Local Urban Plan (*Plan local d'urbanisme PLU*): intercommunally linked with metropolitan region • Municipality Regulatory Land-Use Plan (*Plan local d'urbanisme PLU; Plan de déplacements urbains PDU*): instrument for cities, definition of zoning and traffic regimes • Local Habitat Programme (*Programme local de l'habitat PLH*): strategy for housing	• Urbanism department (*Service de l'Urbanisme*) • Planning Agency (in large cities) (*Agence d'Urbanisme*)

Gender impact on spatial planning

- since 2000 attempts to establish specific gender policies in various cities (e.g. Paris, Rennes, Lormont) can be observed
- rising awareness regarding the topic "women and cities", "gender and cities"
- mostly activist movements and NGOs are concerned with gender; did not yet reach formal urban planning
- Act on Equality between Men and Women, 2014

Gender-related networks

- *Chargées de mission à l'égalité entre les femmes et les hommes*: action on urban planning, transport or safety

Table 2.8 National perspective of Spain

Political framework

- parliamentary democracy and constitutional monarchy
- 17 autonomous regions, two autonomous cities with an extensive degree of autonomy, some greater than others (exceeds that of regions or states in some federal states)

Planning system

- legislative planning power rests with the regions (regional government/parliament)
- central government has only residual competences on spatial planning legislation: content of property rights; valuation of land for eminent domain purposes, national-level transportation infrastructure and environmental legislation
- responsibility for planning resides almost completely within powers and competencies of local authorities, they have the obligation to establish, adopt and revise structure planning and land-use regulations
- municipal plans and their modifications have to be approved by the regional ministry (*Consejería*)
- regional planning is led by three principles: equal distribution of benefits and burdens between the previous landowners when new land is developed; guarantee of proper housing for all citizens and regulated provision of health, education and other services for all citizens; protection of the environment

Levels of spatial planning

Laws/policies on spatial planning	*Planning instruments*	*Organisations and main actors in spatial planning*
National level		
• The Land Act 2007 (*Texto Refundido de la Ley del Régimen del Suelo y Ordenación Urbana*)	• National Plan (*Plan Nacional*) • 5-year housing plans (*Planes de Vivienda*) • Infrastructure Master Plan (*Plan Director de Infrastructuras PDI*)	• Ministry of Development (*Ministerio de Fomento*)
Regional level		
• Regional Land Planning Laws (e.g. in Catalonia, Navarre, Madrid, Basque Country, Valencia) • legislation on protection of rural lands	–	• regional ministry/ministries in charge of territorial planning, transportation and housing (*Consejería*), main tasks: approval of municipal plans, adoption of regional plan, investment in transport infrastructure, regional parks • 17 Territorial Councils (*Comisión Territorial de Urbanismo*)
Provincial level		
–	–	• 52 provincial councils (*Comisión Provincial de Urbanismo*)

Local level		
–	• Land-Use Plans (*planes territoriales*) • municipal ordinances for small municipalities	• 8077 municipal councils (*Ayuntamiento*)

Gender impact on spatial planning

- gender issues are unusual and unknown by practitioners and politicians working on spatial development; more common in scientific circles

National level

- Law for Equality (2007): includes a complete article on urban policies, territorial organisation and housing
- Land Act (2007): in the preliminary heading of Article 2 on the principle of sustainable urban development there is a comment on "equal treatment and opportunities for women and men"
- State Housing Plan (2006) specifically considers women who have been victims of gender violence and single-mother families

Regional level (Catalonia)

- Law for the improvement of neighbourhoods, urban areas and villages (2004): refers to equality of genders in the use of urban space and as a criterion for selecting projects to be funded under the law
- gender-sensitive spatial planning laws in Catalonia and the Basque Country

Local level

- implementation of regional laws: gender has been mostly introduced through training courses and toolkits addressed to policymakers and planners

Specific perspectives: judicial decision strengthens gender-impact assessment

- since 2003 all urban plans have to integrate a gender-impact report in their technical documents; the report did not receive much attention until two judges cancelled two urban municipal plans for not having a gender-impact report
- gender-impact reports are becoming an important element with the potential to be a significant instrument to integrate gender into urban plans

Gender-related networks

- *Grupo de Investigación Génerourban*: academic research and advocacy in the School of Architecture in Madrid, 1999, and in the School of Architecture of Barcelona, 2004
- *La Mujer Construye*; *Colectivo de Mujeres Urbanistas* (Madrid): advocacy and cooperative group of women architects
- *Hiria Kolektiboa* (The Basque Country): advocacy and cooperative group of women architects (defunct)
- *Col·lectiu Punt* 6 (Barcelona), 2008: advocacy and cooperative group of women architects
- UNESCO Chair on Gender, 2016: School of Architecture of Madrid

Table 2.9 National perspective of Greece

Political framework

- centralist state (presidential parliamentary democracy), seven decentralised administrations (*apokentroménes dioikíseis*), 13 administrative regions
- southern welfare state
- Mediterranean syndrome of government: rigid, legalistic and formal models of regulation (La Spina and Sciortino, 1993)

Planning system

- "urbanism tradition": focus on architecture, urban design and townscape
- regulatory planning system with hierarchical structures, strong legalistic tradition, pronounced command and control mechanisms
- focus on control of building permissions, land-use regulation and statutory plans
- besides the rigidity and complexity the system is quite flexible when needed and allows for bypassing laws (informal practices)
- community involvement and consultation mechanisms are underdeveloped

Levels of spatial planning

Laws/policies on spatial planning[a]	*Planning instruments*	*Organisations and main actors in spatial planning*
National level		
• 2508/1997 Sustainable Development of Cities and Settlements (*Viosimi ekistiki anaptixi ton poleon kai ekismon tis choras*) and 2742/1999 Spatial Planning and Sustainable Development (*Chorotaxikos schediasmos kai aeiforos anaptixi*) have been abolished except for a couple of articles • 3894/2010 Acceleration and Transparency in Implementation of "fast-track" Strategic Investments (*Epitachinsi kai diafania ilopiisis stratigikon ependiseon "fast-track"*); 3986/2011 Urgent Measures for the Implementation of the Medium-term Framework of the Financial Strategy 2012–2015 (*Epigonta metra efarmoghis mesoprothesmou plesiou dimosionomikis stratigikis*); 4072/2012 Improvement of the Business Environment (*Veltiosi epichirimatikou perivallontos*): fragmentary planning laws concerning strategic "fast track" investments and public property exploitation • 4269/2014 Spatial and Urban Planning Reform – Sustainable Development (*Chrotaxiki kai poleodomiki metarithmisi – Viosimi Anaptixi*); 4447/2016 Spatial Planning – Sustainable Development (*Chorikos schediasmos – Viosimi anaptixi*): planning reform laws, facilitate exceptional planning and development within the planning system and reinforce strategic and regulatory planning on different levels	• Strategic Spatial Planning/National Spatial Strategy (*Stratigikos Chorikos Schediasmos/Ethniki Choriki Stratigiki*): national territorial objectives across spatial scales and sectors • Special Frameworks for Spatial Planning (*Eidika Chorotaxika Plaisia*): sectoral plans, e.g. energy, land policy	• parliament (legislative power) • ministries (executive power), especially Ministry of the Environment and Energy and Ministry of Infrastructure and Transport • state council (judiciary power) • further institutions: Ministerial Council for the National Spatial Strategy, National Council for Spatial Planning

Third level of administrative division		
–	–	• seven decentralised administrations of the state (third level of administrative divisions) • head: general secretary, appointed by state government
Regional level (second level of administrative division)		
–	• Regional frameworks for spatial planning (*Perifereiaka Plaisia Chorotaxikou Schediasmou*): strategic spatial planning • Regional Spatial Programmes RSP (*Periferiaka Chorika Programmata*)	• 13 administrative regions (elected regional authorities)
Local level (first level of administrative division)		
• 3852/2010 New Architecture of the Local Administration and the Decentralised Administration – *Kallikratis* Programme (*Nea architektoniki tis aftodiikisis kai tis apokentromenis diikisis – programma Kallikratis*): regulates spatial planning on municipality level; delegates power of supervision to regional administrations	• General Spatial Plan GSP (*Genika Chorika Schedia*): general spatial strategy, infrastructure planning and land-use zones for a municipality • Special Spatial Plan SSP (*Eidika Chorika Schedia*): fragmentary plan irrelevant to municipal plans/borders, facilitates focused "fast track" strategic development on public or private land • Town Plan to be Applied TPA (*Poleodomiko Schedio Efarmoghis*): detailed and specific land-use/terms and indices of development; binding for public authorities and private properties	• 325 municipalities, according to Law 3852/2010 • final decision of approval of General Spatial Plan (GSP) and Special Spatial Plan (SSP) are taken by the state/ministry by Presidential Decree, and of Town Plan to be Applied (TPA) by the General Secretary of the decentralised administration, according to law 4447/2016

Gender impact on spatial planning

- no gender policies in the sector of spatial planning and design so far, no relevant overall strategy developed on the national, regional or local level
- significant gap between academic research in spatial planning and design and administrative and professional practices in these fields regarding gender
- Greek crisis deepened gender inequalities
- despite the reconstitution of spatial planning and development processes, gender issues are not on the agenda at the moment
- gender issues in spatial design are partly covered as a by-product in guidance provided on other objectives (i.e. design of playgrounds, inclusive accessibility – design for all)

Specific perspectives: role of court; huge changes in planning policy and system

- council of state plays an important role in planning processes in order to challenge, repeal or validate formally approved plans/administrative decisions
- sometimes appeals to the court serve as a means of participation or social intervention in planning processes
- spatial planning policy and system is currently being transformed (double reform between 2014 and 2016, laws 4269 and 4447), the shift towards more flexible and neoliberal approaches could be seen as one factor why gender is not on the agenda

Note

a Special thanks for the compilation of laws and instruments in the legal planning framework in Greece goes to Dr Charis Christodoulou, Assistant Professor in Urban Planning at the School of Architecture, Aristotle University of Thessaloniki.

Table 2.10 National perspective of Switzerland

Political framework

- half-direct democracy
- three administrative levels: confederation (national state), cantons (26), municipalities
- important principles: federalism, subsidiarity
- federalism: national state grants cantons a high degree of autonomy; cantons grant municipalities a high degree of autonomy
- subsidiarity: social and political issues should be dealt with at the most local level possible
- regional politics are very important because of four different languages and big disparities between metropolitan and rural/mountainous regions

Planning system

- cantons: main planning authorities, coordinate vertical (municipalities – confederation) and horizontal (sectoral) planning, often delegate tasks to the municipalities (local authorities)
- confederation: defines the national framework legislation and sectoral policies; promotes and coordinates the spatial planning of the cantons (26 planning and building laws); is responsible for cross-border or national issues
- municipalities: responsible for practical planning implementation

Levels of spatial planning

Laws/policies on spatial planning	*Planning instruments*	*Organisations and main actors in spatial planning*
National level (confederation/Bundesstaat)		
• Law on Spatial Planning 2015 (*Raumplanungsgesetz*), first implemented in 1969, last amendment in 2015: general planning principles, defines planning hierarchy and instruments • Functional Spatial Planning laws (*Funktionales Raumplanungsrecht*): e.g. Law on Motorways, Law on Nature and Habitat Conservation, etc.	• Sectoral Strategies and Plans (*Sachpläne*), e.g. Sectoral Aviation Infrastructure Plan, binding for cantons, not binding for private individuals • Spatial Concept Switzerland 2012 (*Raumkonzept Schweiz*): first strategic document concerning spatial planning on national level, developed by all three levels of government, not legally binding	• Federal Office for Spatial Development (*Bundesamt für Raumentwicklung ARE/Office fédéral du développement territorial ARE*) • Tripartite (federal/cantonal/municipal) Conference (TK) (*Tripartite Konferenz TK/Conférence Tripartite CT*)
Federal state level (canton/Kanton)		
• 26 different cantonal planning and building laws (*Kantonale Planungs- und Baugesetze*), within the framework (aims and principles as well as instruments) of the federal legislation	• Cantonal Structure Plan (*Kantonaler Richtplan/plan directeur cantonal*): comprehensive planning strategy, covers the whole area of a canton, subject to approval by the Federal Council, guides spatial development, coordinates activities of confederation, canton and municipalities, binding for municipalities, not binding for private persons	• Cantonal Offices for Spatial Development
Regional level		
• cantonal planning and building laws	• Regional Structure Plan (*Regionaler Richtplan/plan directeur régional*): strategic planning instrument • Planning instruments for functional spaces (*funktionale Räume*): e.g. programmes for urban areas (*Agglomerationsprogramme*)	• regional planning associations • supra-municipal planning platforms, e.g.: *Metropolitankonferenz Zürich, Plattform Aargau-Solothurn (PASO), Trinationaler Eurodistrict Basel (TEB)*

Local level (municipality/Gemeinde)		
• Local Building Laws (*Bau- und Zonenordnungen/Loi pour l'aménagement et la construction* LATC)	• Local Structural Plan (*Kommunaler Richtplan, plan directeur communal*): strategic planning instrument • Land-Use Plan (*Nutzungsplan, plan d'utilisation*): delimits building areas from non-building areas, determines type and extent of specific building use, has to meet provisions of Federal Law on Spatial Planning	• municipalities

Gender impact on spatial planning

National level

- no gender-sensitive requirements in Federal Planning Act
- EU laws were taken up by Swiss politics (e.g. Amsterdam Treaty on Gender Mainstreaming 1999)
- gender mainstreaming as political guideline (since 2004 for Swiss federal offices)
- gender equality as a criterion in Swiss monitoring system for sustainable development (MONET)

Cantonal level

- no gender-sensitive requirements in any of the 26 cantonal planning and building laws
- gender is not a specific issue in cantonal structure plans

Local level

- diversity approach is somewhat present in municipal planning
- participatory approaches tend to imply gender/diversity questions
- not a law, but a recommendation by Swiss Society of Engineers and Architects (SIA) norm 112/1 for "Sustainable Building above Ground" focuses on user needs; SIA consultative document 2050 and documentation D 0246 "Sustainable Spatial Development" focus on cooperative planning and fair discourse

Specific perspectives: quality-oriented planning – women and equality issues low on official agendas

- Switzerland lacks institutionalisation of issues of gender and space; gender equality is often considered an additional expense
- gender concerns are – if at all – considered in planning of public space, publicly used space or housing (e.g. gender assessment of the WBS, *Wohnungs-Bewertungs-System*), but not generally
- in *Lares* projects gender issues were part of planning and building projects (esp. buildings for higher education), but only pilot projects; little response when it comes to the explicit implementation of gender issues
- gender experts not obligatory in building and planning committees
- key positions in planning and politics where planning relevant decisions are taken are to a very large extent occupied by men
- women are underrepresented on local, regional, cantonal and federal level committees and in important networks, but if they take leading positions they are recognised equally

Gender-related networks

- *Lares*, association fostering gender equality and supporting women-experts in building and planning processes (funding from the Federal Office for Gender Equality, 2006–2012; since 2012 organised as an association)
- créatrices.ch, "women shape Switzerland", founded in 2017
- *frau und sia*, women network within SIA, since 2004
- P, A, F., "planning, architecture, women", 1994–2012

Insights from the comparison

Stephanie Tuggener, Barbara Zibell and Ulrike Sturm

The aim of comparing these nine countries systematically by the means of factsheets was to find out whether and to what extent gender is an issue in spatial development in each country and how it differs between the countries. We argue that the formalised planning system is a relevant aspect to study in order to find answers to this research question. Comparing and analysing the factsheets we focused on three main aspects:

> What links can be identified between the political framework, the planning system and gender or gender substitute concepts? Are there correlations?
>
> Is gender mentioned in a national planning law or policy? Is gender mentioned in a law or policy or taken into account in a planning instrument on state, federal, regional or local level?
>
> What is the degree of participation? (Possibilities for people to participate in formalised or informal ways[10] are seen as a criterion for a more equal planning culture.)

Table 2.11 shows the results of the comparison in condensed form.

Planning systems, cultures and political systems

The authors observed no obvious correlations between planning systems, planning cultures or political systems with gender approaches. A feasible hypothesis is that states with a federal structure and states with the subsidiarity principle tend to provide more possibilities for public participation. This is, e.g., the case for Austria, Switzerland and Germany as opposed to France or England as more centralist states. In a state with a high level of participation it is more likely that various user needs are incorporated in planning processes – an important concern of gender-sensitive planning. It is obvious that in countries which are going through a crisis of austerity, such as Greece, gender concerns and possibilities to participate are accorded low priority.

Table 2.11 Results of the comparison

Country	*Gender*				*Substitute concepts*			
	Is gender mentioned explicitly in a national planning law?		*Is gender mentioned in a law or taken into account in a planning instrument on national, federal, regional, or local level?*		*Is (social) sustainability mentioned in a national planning law?*		*Degree of participation on political level and in formalised planning procedures* High ● Medium ◑ Low ○	
	Yes	*No*	*Yes*	*No*	*Yes*	*No*	*Political rights/ processes*	*Formal, defined in planning laws*
Finland		•		•	•		●	●
Netherlands		•		•	•[a]		●	●
England		•		•	•		●	◑
Austria		(•)[b]	•			(•)	●	●
Germany	(•) (BauGB)		•		•		●	●
France		•	•		•		●	◑
Spain	•		•		•		●	◑
Greece[c]		•		•	•		◑	◑
Switzerland		•		•	•		●	●

Notes

a The law is currently in transition. The new Environmental Act (*Omgevingswet*) will incorporate sustainability as a leading principle.

b In Austria there is no national planning law. The laws on state level are the basic legislation.

c Greece, under financial control for the last eight years (2010–2018), has limited options for participation in political decisions.

Overall, it can be said that gender is not prominent or important in spatial development in any of the countries we studied, regardless of the political system or the planning system. The analysis showed that gender is rarely institutionalised or part of a formal planning level (neither regarding laws nor policies nor instruments). Male hegemonies are apparently not only related to formal political structures but are deeply rooted in the minds and cultures of different societies; thus, at best, there is only minor awareness of this and it is rather hard to change. In general, it is true for all countries that gender plays a greater role in academia than in the practice of spatial development.

Gender in planning policies

The overview shows that gender is hardly ever mentioned explicitly in a national planning law or policy. Only the national laws of Spain and Germany mention gender, or rather sex, by naming the different needs of women and men. However, it is more likely that gender appears in planning instruments or policies on federal, regional or local level.

In Germany the building code (Baugesetzbuch, BauGB) states that "the social and cultural needs of the population, especially the needs of families, young, old and disabled people, different impacts on men and women ... have to be taken into account whilst developing an urban land use plan" ("Bauleitplan" acc. to §1 (6) Nr. 3). In some regional spatial development concepts (Regionale Raumordnungsprogramme, RROP) or plans on federal state level, gender equality or gender-sensitive spatial planning are explicit objectives (e.g. RROP Hannover, Stuttgart Regional Plan, Ruhr Regional Plan, North Rhine-Westphalia Regional Development Plan). In the process of developing the Ruhr Regional Plan (still ongoing) gender objectives and gender expertise were considered very important (Bauer and Frölich v. Bodelschwingh, 2017). Women-networks such as Frauennetzwerk Ruhrgebiet played an important role in the definition of these objectives. Important bases to assess the development of gender equality are the gender-sensitive statistics as introduced in Germany by the project "Frauen Männer Räume" (Women men spaces) (BBR, 2007).

In Spain gender equality is part of the national law (Land Act) as well as some regional laws. Uniquely in comparison with all other countries we studied, in Spain all urban plans have had to incorporate a Gender Impact Report since 2003.

In Austria planning is not regulated on national level but on the level of the federal states. Thus, the role of gender issues in the different federal states differs greatly. Vienna as a city with federal status leads the way. With its long experience, many publications and pilot projects (e.g. the handbook *Gender Mainstreaming in Urban Planning and Urban Development*, Gender-Leitprojekt Seestadt Aspern) it has a unique position in the European context. Certain cities in Germany such as Berlin, Munich and Freiburg sometimes implement gender-sensitive methods. Best-practice examples are, e.g., the gender assessment ("Gender-Check") of the local transportation plan in the City of Berlin, the city development concept "Perspective Munich" with the involvement of gender expertise in the course of the evaluation (LH München, 2008) or the incorporation of gender criteria in the process of planning the light rail's extension in the City of Freiburg (Bauer and Frölich v. Bodelschwingh, 2017).

In Switzerland gender aspects are not included in any law or planning instrument. However, the national office for gender equality has supported pilot projects such as "Frau am Bau" (Women in construction, 1997–2003) and "Lares" (2006–2012) which aimed to enhance gender-sensitive and everyday-needs-specific construction and planning. In pilot projects "Lares" ensured that gender know-how is taken into account by developing gender criteria, carefully accompanying the processes and introducing a gender perspective on the decision-making level (see Chapter 5). Since 2012 "Lares" has been organised as an association and no longer receives national funding. The goals set up by "Lares" are supported by objectives for social sustainability, e.g. in the guidelines of the Swiss association of engineers and architects SIA "Standard guide to SIA norm 112 – Sustainable construction – Structural engineering" (SIA, 2005) and "Leaflet 2050 and documentation 0246. Sustainable spatial development – Local and regional planning" (SIA, 2015).

On the other hand, there are countries where the influence of gender in spatial development appears mostly bottom-up – if at all. In the Netherlands and in France there are various bottom-up groups which claim the importance of gender in spatial development. Common issues they address are housing, safety in public spaces, the right to the city or the need to consult the experience of various user groups. As an overarching goal, gender-oriented bottom-up initiatives aim to promote a planning culture that is gender equal, that takes the daily realities of all people undertaking housework and care work into account and that promotes the equality of women within planning professions.

A peculiarity nevertheless is Finland's "gender blindness". From a European perspective it would be assumed that Finland as a Scandinavian country is especially gender-sensitive. The International Gender Equality Prize that was launched by the Finnish Government in 2017 underlines this assumption (e.g. International Gender Equality Prize 2018). However, according to Finnish experts, gender as a concept or category is not taken into account at all in spatial development. Apparently, internal and external perspectives on the importance of gender equality in Finland differ greatly.

Whether and how the activities of bottom-up groups reach mainstream, institutionalised planning is a subject for further research. There is a need to critically reflect on how gender relations in the work of bottom-up groups seem to continue with hegemonic male and subordinate female roles (for activist movements see also Chapter 6). Moreover, it would be interesting to analyse whether there are differences between cities – where there are usually many activist movements – and the countryside. Of course, bottom-up groups also exist in countries where gender issues are to some extent institutionalised. Bottom-up and institutionalised practices are not mutually exclusive but can enrich one another.

Furthermore, we observed that (broader) substitute concepts such as social sustainability or integrated approaches that weigh the three pillars of sustainability equally, as stated in the Leipzig Charter on Sustainable European Cities, have become part of planning laws since the 1990s. According to the Leipzig Charter "all dimensions of sustainable development should be taken into account at the same time and with the same weight" (EU, 2007: 1). This is the case for all countries considered in the comparison. Other guiding principles which emerged from feminist claims have been adopted by the mainstream, for instance the concept of "the city of short distances". On an international level, political concepts currently include gender topics as a matter of course (see e.g. Sustainable Development Goals). However, as soon as a feminist concept is incorporated by the mainstream its origin is often no longer acknowledged. As already mentioned in the section on utopian feminists and social movements (see above), both in the past and to date, neither feminist architectural progress nor feminist urban development concepts are recognised as being of feminist origin if taken over by the mainstream.

Overall, it can be concluded that gender concerns such as the consideration of housework and care work or unpaid work, the gendered division of labour, gendered hierarchies in the labour market and social roles are hardly ever taken into account by planning laws or instruments on any spatial level. Moreover, we observed that gender competence is often missing in planning institutions. If there is gender expertise in a government, it is usually on the national or federal state level (e.g. Federal Office for Gender Equality in Switzerland or Gender Equality Unit in Finland) and focuses on "non-spatial" topics such as the labour market in general or education, or simply ignores the spatial dimension.

Participation as a criterion of gender-sensitive planning

Participation in its various forms principally offers ways of bringing in the perspectives of different interest and user groups. *Formal participation* as regulated according to spatial planning laws and policies for the public and for interest groups exists in all nine countries. Thus, the public has to be actively informed about the objectives and procedures of spatial planning. This can take the form of an initial consultation. Often, interest groups are officially involved in the design process of spatial development concepts. Land-use legislation in practically all countries includes procedures for public participation in the adoption processes: drafts of plans are publicly disclosed, suggestions and objections can be placed during a period of public display, and responses have to be published. Once a plan is adopted, citizens may appeal to administrative courts to challenge it. In accordance with the legal systems, it is possible to claim one's legal rights. Through its institutionalisation, formal participation may act in a corrective manner, yet it is difficult to place issues if they are not promoted by strong stakeholder groups. Gender issues are very often not formalised enough to have a bearing. Moreover, the standard reached in post-war planning laws or policies is threatened under economic pressure. European austerity policies lead

– among other things – to new "simplified" planning procedures at the cost of democratic standards. Thus, in Greece the formerly obligatory participatory processes have been erased from the legal framework of plan preparation in the recent planning reform law (4269/2014, 4447/2016; see also Chapter 8).

At the local level in many countries, the design of planning and development concepts tends to be more people-oriented than on national or federal level. To name a few: in Austria, many of the communities, especially for the preparation of the local development programme, establish their own participation processes, although this is not obligatory. In England, groups of local people can work out a plan for their own neighbourhood, a Neighbourhood Plan, which has to be in accordance with national policy and other local plans. "Municipal Committees of Consultation" (Law 3852/2010) are entrusted with the public participation procedure for local level plans in Greece, but they do not function in all municipalities. Of late, some cities in Spain have started developing wider mechanisms for the participation of citizens in decision-making processes on public investments in small neighbourhood projects. These activities are often a supplement to formal planning and not obligatory.

Formal participation is amended by a type of informal participation which is neither obligatory nor follows specific rules as defined in planning laws, policies or local regulations. In many countries, this type of informal participation was inaugurated in the 1960s and 1970s by social movements, attracting less attention during the following decade and regaining momentum since the 1990s. Informal participatory processes have been promoted within spatial development as part of politically established programmes supporting civil society actions since the 1990s, when it became clear that complex planning challenges profit from the involvement of a variety of stakeholders, as in Agenda 21 (United Nations, 1992) and the following processes (see Chapter 7). Here gender topics became part of the promotion of sustainability.

In general, it can be said that gender issues may be part of formal or informal participation processes, but this is not necessarily the case. Whether or not gender issues are brought forth depends on the details of the process. The success of informal participation depends greatly on the stability and influence of civil society institutions and on an open political attitude and awareness. Interests and power relations influence the ability of different social groups (including gender aspects) to assert their requests. Informal participatory processes may even lead to a backlash when the turn from rationalist to communicative planning does not empower "unheard voices" but becomes a means to stabilise power relations and rather reinforces inequalities instead of reducing them (see e.g. Beebeejaun, 2006).

Principally, participation, whether formalised or informal, has to be understood in the context of power relations (Healey, 2003). Explicit gender strategies, however, tend to use participation as a means of "empowerment" and enhance the democratic quality of planning by involving various interest and user groups (e.g. Young, 2005; Urbact, 2006). Self-organisation as a bottom-up strategy is a way of bringing forth gender concerns – some groups make use of digitalisation to gain power (see Chapter 6), amending formal processes or precisely when formal approaches do not have good prospects (for an overview with an international perspective see Beebeejaun, 2016).

The history of gender in spatial planning and the current situation in nine European countries show a mixed picture. There has been a lot of progress regarding gender theory, and the different waves of women's movements achieved important advances. Nevertheless, in all the countries gender is far from being institutionalised on a national level and no planning system is especially gender-sensitive. However, there are pilot projects and promising practices, mostly on a city level where gender issues are implemented in planning processes despite the challenges and contradictions that continue to exist. Chapter 3 will explicitly deal with contradictions and challenges that arise in the context of the gender debate and the implementation of gender concepts.

Notes

1 In the book *Zur Entwicklung der Stadtplanung in Europa* (Albers, 1997), Gerd Albers, a practical as well as academically experienced and well-known international cross-linked German architect and urbanist (named also the "pope of urbanism"), until 1987 Professor for Urbanism and Regional Planning at the Technical University of Munich, describes the processes of upcoming town planning in different European countries since the nineteenth century, among them also the nine referenced to here in this book. The summarising conclusions within the following section are based on this publication.

2 Verein Margarete Schütte-Lihotzky Club: Association to discuss and collect the work of Margarete Schütte-Lihotzky located in Vienna, Austria; www.schuette-lithotzky.at.

3 Feministische Organisation von Planerinnen und Architektinnen (FOPA) – feminist organisation of female planners and architects in Germany since 1981. From 1983 to 2004, FOPA published a series called "Freiräume", which provided ongoing information on feminist theories and practice in planning, architecture and construction (c.f. www.fopa.de).

4 *Zoll+* (formerly *zolltexte*) is the Austrian Magazine of Landscape and Open Space. It is a platform for the constructive discussion of planning topics with a focus on urban and rural landscapes. *Zoll+* appears biannually in an edition of about 1,500 pieces. *Zoll+* has been edited by Forum Landschaftsplanung since 1991 and is based on voluntary work (see www.foruml.at/, 15 April 2016).

5 http://habitat.aq.upm.es/boletin/n7/acharter.html – The European Charter for Women in the City was the product of action-oriented research co-financed by the European Commission – Directorate-General (DGV) for Employment, Industrial Relations and Social Affairs. https://femmesvilles.org/: Women in Cities International (WICI) was founded in 2002 as a not-for-profit network focusing on gender equality and the participation of women in urban development. It was incorporated in June 2003. Based in Montréal, WICI grew out of work undertaken in Canadian cities in the 1990s on women's safety in urban settings, and the participatory potential of women's safety audits to empower women and create dialogue and change within city governments. http://womenfriendlycities.com: "Women Friendly Cities United Nations Joint Programme" was initiated in 2006 with the aim of mainstreaming gender equality in the planning and programming processes of local administrations. In parallel with this process, the Women Friendly Cities, United Nations Joint Project (UNJP) aimed at empowering local administrations and women's organisations to establish strong collaboration among them.

6 The Bundesamt für Bauwesen und Raumordnung (German Federal Office for Building and Regional Planning – BBR), academic and political consultancy for the German Federal Government or its predecessor institution, the Bundesforschungsanstalt für Landeskunde und Raumordnung (Federal Research Institute for Regional Studies and Planning – BfLR) focused on the perspectives of women on regional research and regional spatial planning as early as the beginning of the 1990s (c.f. *IzR – Informationen zur Raumentwicklung* – Information on Spatial Regional Development, Volume 8/9.1990, *Materialien zur Raumentwicklung* – Material on Spatial Regional Development, Volume 38.1991). A colloquium on the theme of "Frauen planen die Stadt" (Women plan the city), held by the BMBau (Federal Building Ministry) and BfLR in 1993 in Bonn, was published in the research series of the Federal Building Ministry of the time (Volume 493.1993). Another volume on the subject of "Frauen in Europa" (Women in Europe) with comparative perspectives on Finland, the Netherlands and the Federal Republic of Germany, together with summaries in English, French and Italian, appeared in 1995 (IzR 1.1995).

7 For example, in Austria, Germany, Finland, Great Britain and Spain (see also Chapter 3, Box 3.3).

8 The European Spatial Development Perspective (ESDP) is a document approved by the Informal Council of Ministers of Spatial Planning of the European Commission in Potsdam in 1999. It is a legally non-binding document forming a policy framework with 60 policy options for all tiers of administration with a planning responsibility. The strategic aim is to achieve a balanced and sustainable spatial development strategy.

9 For references to the country tables see below. The national factsheets are to a large extent based on the knowledge of the members of the working group.

10 When looking at spatial planning and development in Europe, it is possible to distinguish between formal and informal participation practices. Formal participation is guaranteed by law and has been part of national planning laws or policies since the 1950s and 1960s. Additionally, informal participation practices have developed over recent decades. Informal participation can be aligned with formal processes and an officially accepted or even adopted supplement to them; on the other hand, informal participation can also arise from social movements to raise voices that would otherwise remain unheard.

References

Albers, G. (1997). *Zur Entwicklung der Stadtplanung in Europa. Begegnungen, Einflüsse, Verflechtungen (The development of urban planning in Europe. Encounters, influences, interdependencies)*. Bauwelt Fundamente 117. Braunschweig/Wiesbaden: Vieweg.

Allmendinger, P. (2002). *Planning theory*. New York: Palgrave.

Altenstraßer, C., Hauch, G. and Kepplinger, H. (Eds.) (2007). *Gender housing. Geschlechtergerechtes bauen, wohnen, leben (Gender housing. Gender-fair building, dwelling, living)*. Innsbruck, Wien, Bozen: StudienVerlag.

Aufhauser, E., Herzog, S., Hinterleitner, V., Oedl-Wieser, T. and Reisinger E. (2003). *Grundlagen für eine Gleichstellungsorientierte Regionalentwicklung (Basics studies for an equality oriented regional development)*. Studie im Auftrag des Bundeskanzleramtes. Abteilung IV/4. Wien.

Bacchi, C. (2011). Gender mainstreaming and reflexivity. Asking some hard questions. In: *Advancing Gender+ Training in Theory and Practice Conference*. An International Conference for Practitioners, Experts and Commissioners in Gender+ Training, 3–4 February 2011, Madrid, 30–45.

Bate, M. (2018). *The periodic table of feminism*. London: Penguin Random House, UK.

Bauer, U. and Frölich v. Bodelschwingh, F. (2017). *30 Jahre Gender in der Stadt- und Regionalentwicklung (30 years of gender in city and regional development)*. Berlin: Deutsches Institut für Urbanistik. Available at: https://difu.de/publikationen/2017/30-jahre-gender-in-der-stadt-und-regionalentwicklung.html [Accessed 9 Aug. 2018].

Bauhardt, C. (1995). *Stadtentwicklung und Verkehrspolitik. Eine Analyse aus feministischer Sicht (Urban development and transport policies. An analysis from a feminist perspective)*. Stadtforschung aktuell. Basel: Birkhäuser.

Baumeister, R. (1876). *Stadterweiterungen in technischer, baupolizeilicher und wirtschaftlicher Beziehung (City expansion in a technical, building policy and economic context)*. Berlin: Ernst & Korn.

Baumgart, S. and Seggern, H. von (1996). *Frauengerechte Stadtplanung – ein Beitrag zu einer "gender-sensitive"-Planung der Stadt (Women-friendly urban planning – a contribution to a gender-sensitive planning of the city)*. Schriftenreihe Forschung des Bundesministeriums für Raumordnung, Bauwesen und Städtebau, Heft 498. Bonn.

Bauriedl, S., Schier, M. and Strüver, A. (Eds.) (2010). *Geschlechterverhältnisse, Raumstrukturen, Ortsbeziehungen: Erkundungen von Vielfalt und Differenz im spatial turn (Gender relation, structure of space and spatial relations. Investigations of diversity and difference in spatial turn)*. Reihe Forum Frauen- und Geschlechterforschung. Münster: Verlag Westfälisches Dampfboot.

Bauwelt 31/32 (1979). *Frauen in Architektur – Frauenarchitektur (Women in architecture – women's architecture)*. Edition 79.

BBR – Bundesamt für Bauwesen und Raumordnung (Ed.) (2007). *Frauen – Männer – Räume (Women – men – spaces)*. Bd. 26. Bonn.

BBR – Bundesamt für Bauwesen und Raumordnung (Ed.) (2006). *Gender Mainstreaming im Städtebau (Gender mainstreaming in urbanism)*. ExWoSt-Informationen, 26/5-04/2006. [Online] Bonn. Available at: www.bbsr.bund.de/BBSR/DE/Veroeffentlichungen/ExWoSt/24_29/exwost26.html [Accessed 23 May 2018].

Beauvoir de, S. (1949). *Le Deuxième Sexe (The second sex)*. Paris: Gallimard.

Becker, R. (2009). *Frauenwohnprojekte – keine Utopie! Ein Leitfaden zur Entwicklung autonomer Frauen(wohn)räume mit einer Dokumentation realisierter Projekte in Deutschland (Women's housing projects – no utopia! A guideline for the development of autonomous women's (living) spaces with documentation of realised projects in Germany)*. Studien Netzwerk Frauenforschung NRW Nr. 3, Dortmund.

Becker, R. (2008). Angsträume oder Frauenräume? Gedanken über den Zugang von Frauen zum öffentlichen Raum (Places of fear or women's spaces? Thoughts about women's access to public space). In: Feministisches Kollektiv (Ed.), *Street Harassment. Machtprozesse und Raumproduktion (Street harassment, power relation and production of space)*. Buchreihe der ÖH Uni Wien, Bd. 3. Wien: Mandelbaum Verlag, 56–74.

Becker, R. (1997). Frauenforschung in der Raumplanung – Versuch einer Standortbestimmung (Women's studies in spatial planning – attempt to determine the current situation). In: C. Bauhardt and R. Becker (Eds.), *Durch die Wand! Feministische Konzepte zur Raumentwicklung (Through the wall! Feminist concepts for spatial development)*. Pfaffenweiler: Centaurus, 11–32.

Becker, R. and Kortendiek, B. (Eds.) (2004). *Handbuch Frauen- und Geschlechterforschung: Theorie, Methoden, Empirie (Handbook of women's and gender studies. Theory, methods and empiricism)*. Reihe "Geschlecht & Gesellschaft", Bd. 35, 1st edn. Wiesbaden: VS Verlag für Sozialwissenschaften.

Beebeejaun, Y. (Ed.) (2016). *The participatory city*. Berlin: Jovis.

Beebeejaun, Y. (2006). The participation trap. The limitation of participation for ethnic and racial groups. *International Planning Studies*, 11(1), 3–18.

Benevolo, L. (1983). *Die Geschichte der Stadt (The history of the city)*. Frankfurt/New York: Campus.

Bock, S., Heeg, S. and Rodenstein, M. (1997). Reproduktionskrise und Stadtstruktur. Zur Entwicklung von Agglomerationsräumen aus feministischer Sicht (Reproduction crisis and urban structure. On the development of agglomerations from a feminist perspective). In: C. Bauhardt and R. Becker (Eds.), *Durch die Wand! Feministische Konzepte zur Raumentwicklung (Through the wall! Feminist concepts for spatial development)*. Pfaffenweiler: Centaurus, 33–52.

Booth, C. and Bennett, C. (2002). Gender mainstreaming in the European Union. Towards a new conception and practice of equal opportunities? *The European Journal of Women's Studies*, 9(4), 430–446.

Bradshaw, S. (2015). Engendering development and disasters. *Disasters*, 39(s1), s54–s75.

Braun, L. (1901). *Frauenarbeit und Hauswirthschaft (Women's work/Female labour and domestic economy)*. Berlin: Vorwärts.

Buchmüller, L. (1995). Ausradierte Geschichte. Dohna-Poninska und die Anfänge der Stadtplanungstheorie (Eradicated history. Dohna-Poninska and the beginnings of city planning theory). *disP – The Planning Review*, 31(120), 11–17.

Busch-Sperveslage, A. (1999). *Die Klosterkirche in Wennigsen (The monastery church in Wennigsen)*. Studien zur Bauforschung Nr. 18, Stuttgart, Koldewey-Gesellschaft (Ed.) (Diss. Universität Hannover 1998).

Bustelo, M. (2017). Evaluation from a gender+ perspective as a key element for (re)gendering the policymaking process. *Journal of Women, Politics and Policy*, 38(1), 84–101.

Butler, J. (1991). *Das Unbehagen der Geschlechter (Gender trouble)*. Frankfurt am Main: Suhrkamp.

Campbell, S. and Fainstein, S. (Eds.) (1996). *Readings in planning theory*. Oxford: Wiley-Blackwell.

Dame, T. and Grant, A. (2001). *Kelowna Planning for Safer Communities Workshop Report*. Cowichan Valley Safer Futures Program, Canada.

Damyanovic, D. (2013). Gender mainstreaming as a strategy for sustainable urban planning procedures. In: I. Sánchez de Madariaga and M. Roberts (Eds.), *Fair shared cities. The impact of gender planning in Europe*. Farnham: Ashgate, 177–199.

Damyanovic, D. (2007). *Landschaftsplanung als Qualitätssicherung zur Umsetzung der Strategie des Gender Mainstreaming (Landscape planning as quality assurance for the implementation of the strategy of gender mainstreaming)*. Vienna: Guthmann-Peterson.

Damyanovic, D. and Fuchs, B. (2014). Gender studies in planning processes – examples from Austria. In: *8th European Conference on Gender Equality in Higher Education*. Book of Abstracts. Vienna: Office for Gender Competence, Vienna University of Technology.

Damyanovic, D. and Zibell, B. (2013). Is there still gender on the agenda for spatial planning theories? Attempt to an integrative approach to generate gender-sensitive planning theories. *disP – The Planning Review*, 49(4), 25–36.

Diesenreiter, C. (2015). *Freiräume in selbstbestimmten Frauenwohnprojekten. Vergleichende Analyse von Frauenwohnprojekten in Wien und Berlin und deren Beitrag zu aktuellen Planungsdebatten im Kontext einer alltagstauglichen und partizipativen Freiraumplanung im Wohnbau (Open spaces in self-determined women's residential projects. Comparative analysis of women's residential projects in Vienna and Berlin and their contribution to current planning debates in the context of participatory open space planning suitable for everyday use in housing)*. Master's thesis. Universität für Bodenkultur, Vienna.

Doan, P.L. (Ed.) (2015). *Planning and LGBTQ communities. The need for inclusive queer spaces*. London: Routledge.

Döge, P. (2001). *Geschlechterdemokratie als Männlichkeitskritik: Blockaden und Perspektiven einer Neugestaltung des Geschlechterverhältnisses (Gender democracy as criticism on masculinities. Obstacles and perspectives of the construction of gender relations)*. Wissenschaftliche Reihe. Bielefeld: Kleine.

Dohm, H. (1874). *Die wissenschaftliche Emancipation der Frau (The academic emancipation of women)*. Berlin: Wedekind & Schwieger.

Dohna-Poninska von, A. alias Arminius (1874). *Die Grossstädte in ihrer Wohnungsnoth und die Grundlagen einer durchgreifenden Abhilfe (The big cities in their housing shortage and the basics of a drastic remedy)*. Leipzig: Duncker & Humblot.

Dörhöfer, K. (1994). Raumplanung für Frauen: Theoretische Positionen und praktische Konzepte in Deutschland (Spatial planning for women. Theoretical positions and practical concepts in Germany). *Perspektiven*, 1/2, 19–23.

Dörhöfer, K. (1990). Feministische Ansätze in der Architekturausbildung (Feminist approaches in architectural training). In: H. Burmester, G. Cissek, C. Ernst, D. Ewald, E. Helwes, Nanny, H. Nieselt, E. Stoll and B. Weißer (Eds.), *Feministische Ansätze in der Architekturlehre, Symposion – Documentation of Frauen-AG Architektur HdK, TU Berlin 6.2.1989 (Feminist approaches in studies of architecture, Symposion – Documentation of Women's Working Group Architecture HdK University of the Arts, TU Technical University of Berlin)*, 6–25.

Dörhöfer, K.; Flecken, U. and Zibell, B. (1998): Frauenforschung und Gender Studies in der Stadt- und Regionalplanung. (Women´s and gender studies in urban and regional planning). In: *Stadtplanung und Städtebau. Positionen finden – Überzeugungen vermitteln. (Urban planning and urbanism. Finding positions – Conveying convictions)*. Hg. Vereinigung für Stadt-, Regional- und Landesplanung e.V. (SRL), Berlin: SRL, 43–57.

Dörhöfer, K. and Terlinden, U. (1998). *Verortungen: Geschlechterverhältnisse und Raumstrukturen (Contextualisations. Gender relations and spatial structures)*. Basel: Birkhäuser.

EU – European Union (2007). *Leipzig Charter on Sustainable European Cities. Final Draft (2 May 2007)*. [Online]. Available at: http://ec.europa.eu/regional_policy/archive/themes/urban/leipzig_charter.pdf [Accessed 4 Aug. 2018].

European Commission (2014). *Report on Equality of Women and Men 2014*. Luxembourg.

Fainstein, S. and Servon, L. (Eds.) (2005). *Gender and planning. A reader*. New Brunswick, NJ: Rutgers University Press.

Fenster, T. (2005). The right to the gendered city. Different formations of belonging in everyday life. *Journal of Gender Studies*, 14(3), 217–231.

Forum Landschaftsplanung (Ed.) (2007). *gender_n. Zoll+*. Österreichische Schriftenreihe für Landschaft und Freiraum, Bd. 10.

Forum Landschaftsplanung (Ed.) (1993). *Planerinnen* (Female planning experts). *zolltexte*. Bd. 2.

Frauenlobby Städtebau (Ed.) (1993). *Frau Stadt Angst Raum. Wie frei bewegen sich Zürichs Frauen in ihrer Stadt? (Woman city intimidating space. How uncontrolled do the women of Zurich move within their town?)*. Zürich: self-published.

Frey, K. and Perotti, E. (2015). *Theoretikerinnen des Städtebaus. Texte und Projekte für die Stadt (Women theoreticians in urbanism. Words and projects for the city)*. Berlin: Dietrich Reimer Verlag.

Gerhard, U. (2012). *Frauenbewegung und Feminismus. Eine Geschichte seit 1789 (The women's movement and feminism. A history since 1789)*. Munich: Beck.

Gildemeister, R. (2017). Doing Gender. Eine mikrotheoretische Annäherung an die Kategorie Geschlecht (Doing gender. A micro-theoretical approach to the category of gender). In: B. Kortendiek, B. Riegraf and K. Sabisch (Eds.), *Handbuch Interdisziplinäre Geschlechterforschung (Handbook of interdisciplinary gender research)*, Reihe "Geschlecht und Gesellschaft", Bd. 65. Wiesbaden: Springer VS. [Online]. Available at: https://doi.org/10.1007/978-3-658-12500-4_35-1. [Accessed 31 Oct. 2018].

Gildemeister, R. (2004). Doing Gender. Soziale Praktiken der Geschlechterunterscheidung (Doing gender. Social practices of gender discrimination.) In: R. Becker and B. Kortendiek (Eds.), *Handbuch Frauen- und Geschlechterforschung. Theorie, Methoden, Empirie (Handbook of women's and gender studies. Theory, methods, empiricism)*, Reihe "Geschlecht & Gesellschaft", Bd. 35, 1st edn. Wiesbaden: VS Verlag für Sozialwissenschaften, 132–140.

Greed, C. (2005). Overcoming the factors inhibiting the mainstreaming of gender into spatial planning policy in the United Kingdom. *Urban Studies*, 42, 719–749.

Greed, C. (1994). *Women and planning. Creating gendered realities*. London: Routledge.

Grüger, C. (2000). *Nachhaltige Raumentwicklung und Gender Planning. Das Beispiel der Regionalplanung beim Verband Region Stuttgart (Sustainable spatial development and gender planning. Example of regional planning by the Union Region Stuttgart, Germany)*. Dortmunder Beiträge zur Raumplanung Blaue Reihe. IRPUD, Dortmund.

Hayden, D. (1981). *The grand domestic revolution. A history of feminist approach for American homes, neighbourhoods and cities*. Cambridge, MA: MIT Press.

Hayden, D. (1977). *Seven American utopias. The architecture of communitarian socialism 1790–1975*. Cambridge, MA: MIT Press.

Healey, P. (2003). Collaborative planning in perspective. *Planning Theory*, 2(2), 101–123.

Hiort, K. and Zibell, B. (1980). *Frauen in Wohnumfeld und Nachbarschaft – Auswirkungen von Bau- und Nutzungsstrukturen auf das nachbarschaftliche Verhalten von Frauen. Untersuchung am Beispiel eines innerstädtischen Altbauquartiers (Berlin-Moabit). (Women in living environment and neighbourhood – effects of building and utilisation patterns on the neighbourly behaviour of women. Study using the example of an inner-city old building district (Berlin-Moabit))*. Diploma thesis at the Faculty of Societal and Planning Sciences of the Technical University of Berlin. Published by Zibell, B. in 1983.

Hofmeister, S., Katz, C. and Mölders, T. (2013). Grundlegungen im Themenfeld Geschlechterverhältnisse und Nachhaltigkeit (Foundations in the thematic field of gender relations and sustainability). In: S. Hofmeister, C. Katz and T. Mölders (Eds.), *Geschlechterverhältnisse und Nachhaltigkeit. Die Kategorie Geschlecht in den Nachhaltigkeitswissenschaften (Gender relations and sustainability. The category gender in sustainability sciences)*. Opladen, Berlin and Toronto: Verlag Barbara Budrich, 33–76.

Horelli, L. (2017). Engendering urban planning in different contexts – successes, constraints and consequences. *European Planning Studies*, 25(10), 1779–1796.

Horelli, L., Booth, C. and Gilroy, R. (2000). *The EuroFEM toolkit for mobilizing women into local and regional development*. [Online]. Helsinki: Helsinki University of Technology, Centre for Urban and Regional Studies. Available at: https://issuu.com/eva_alvarez/docs/12_eurofem_toolbox [Accessed 4 Aug. 2018].

Huning, S. and Mölders, T. (2017). "Gender" in der Lehre für die Stadt von morgen. Ein widerspenstiges Konzept? ("Gender" in teaching for the city of tomorrow. A rebellious concept?) *Feministische Geo-RundMail*, 72, 13–17.

Jarvis, H., Kantor, P. and Cloke, J. (2009). *Cities and gender*. London: Routledge.

Kail, E. and Irschik, E. (2013). Vienna. Progress towards a fair shared city. In: I. Sánchez de Madariaga and M. Roberts (Eds.), *Fair shared cities. The impact of gender planning in Europe*. Farnham: Ashgate, 193–230.

Kail, E. and Kleedorfer, J. (Eds.) (1991). *Wem gehört der öffentliche Raum. Frauenalltag in der Stadt (Who owns the public space. Women's everyday life in the city)*. Wien Köln Weimar: Böhlau Verlag.

Karácsony, M. and Zibell, B. (Eds.) (2018). *Frauennetzwerke in Architektur & Planung. Erfahrungen Orientierungen (Women's networks in architecture & planning. Experiences orientations)*. Zürich: Edition Hochparterre.

Knapp, G.-A. and Wetterer, A. (Eds.) (2007). *Achsen der Differenz. Gesellschaftstheorie und Feministische Kritik (Axes of difference. Social theory and feminist critique)*. Münster: Verlag Westfälisches Dampfboot.

Knoll, B. (2008). *Gender Planning. Grundlagen für Verkehrs- und Mobilitätserhebungen (Gender planning. Principles of traffic and mobility surveys)*. Saarbrücken: vdm Verlag Dr. Müller.

Kortendiek, B., Riegraf, B. and Sabisch, K. (Eds.) (2019). *Handbuch Interdisziplinäre Geschlechterforschung (Handbook of interdisciplinary gender research)*. Reihe "Geschlecht und Gesellschaft", Bd. 65. Wiesbaden: Springer VS.

Kose, U. and Lička, L. (1993). Frage und Antwort. Sind Frauen anders? (Questions and answers. Are women different?). *zolltexte*, 3, 38–41.

Kramer, C. (Ed.) (2002). *FREI-Räume und FREI-Zeiten. Raum-Nutzung und Zeit-Verwendung im Geschlechterverhältnis (FREE spaces and FREE times. Use of space and time in matters of gender relations)*. Schriften des Heidelberger Instituts für Interdisziplinäre Frauen- und Geschlechterforschung (HIFI) e.V. Baden-Baden: Nomos Verlagsgesellschaft.

Kuhlmann, D. (2002). *Raum, Macht und Differenz. Genderstudien in der Architektur (Space, power and difference. Gender studies in architecture)*. Vienna: Luftschacht.

Kuhn, G. (1998). *Wohnkultur und kommunale Wohnungspolitik in Frankfurt am Main 1880 bis 1930 (Housing culture and municipal housing policy in Frankfurt am Main 1880 to 1930)*. Dissertation. Technical University Berlin. Bonn.

Lada, A.S. (Ed.) (2009). *Teaching gender, diversity and urban space. An intersectional approach between gender studies and spatial disciplines*. A book series by ATHENA. Utrecht: Zuidam Uithof Drukkerijen.

LH München – Landeshauptstadt München, Referat für Stadtplanung und Bauordnung (2008). *PERSPEKTIVE MÜNCHEN. Evaluierungsbericht 2007 – Kurzfassung (Perspective for Munich. Report of Evaluation 2007 – Executive Summary)*. München. [Online]. Available at: www.muenchen.de/rathaus/Stadtverwaltung/Referat-fuer-Stadtplanung-und-Bauordnung/Stadtentwicklung/Perspektive-Muenchen/Publikationen.html [Accessed 30 Oct. 2018].

Lihotzky, G. (1927). Rationalisierung im Haushalt (Rationalisation in the household). *Das Neue Frankfurt. Monatsschrift für die Frage der Grossstadt-Gestaltung*, 5, 120–123.

Miko, K., Neureiter, P. and Stadler-Vida, M. (2012). *Planen – aber sicher! Physische und soziale Verunsicherungsphänomene – Wie kann die Stadtplanung ihnen begegnen? (Safe planning! Phenomena of physical and social insecurity – what can urban planning do to help?)*. Werkstattbericht, Nr. 125. Vienna: Stadtentwicklung Wien – Magistratsabteilung 18 (MA 18), Stadtentwicklung und Stadtplanung (Urban Development Vienna – Municipal Department 18, Urban Development and Planning). Available at: www.wien.gv.at/stadtentwicklung/studien/pdf//b008269.pdf [Accessed 15 Nov. 2018].

Mölders, T., Kanning, H. and Zibell, B. (2014). Energiewende im Raum – Impulse der Genderforschung (Energy transition related to space – impulses from gender studies). *Unimagazin. Forschungsmagazin der Leibniz Universität Hannover*, (3/4), 18–22.

Moser, C. and Levy, C. (1986). *A theory and methodology of gender planning. Meeting women's practical and strategic needs*. DPU Gender and Planning Working Paper No. 11, Development Planning Unit, Bartlett School of Architecture and Planning University College London.

Nadin, V. and Stead, D. (2008). European spatial planning systems social models and learning. *disP – The Planning Review*, 44(172), 35–47.

Oedl-Wieser, T. (2006). *Frauen und Politik am Land (Women and politics in rural areas)*. Forschungsbericht, No. 65. Bundesanstalt für Bergbauernfragen, Vienna.

Oswin, N. (2008). Critical geographies and the uses of sexuality. Deconstructing queer space. *Progress in Human Geography*, 32(1), 89–103.

Pascall, G. and Lewis, J. (2004). Emerging gender regimes and policies for gender equality in a wider Europe. *Journal of Social Politics*, 33, 373–394.

Penny, L. (2018). *LAURIE PENNY DOT COM*. [Online]. Available at: https://laurie-penny.com/ [Accessed 2 July 2018].

Pirhofer, G. (1978). Linien einer Kulturpolitischen Auseinandersetzung in der Geschichte des Wiener Arbeiterwohnungsbaues (Lines of a cultural-political conflict in the history of Viennese construction of worker housing). *Wiener Geschichtsblätter*, 1.

Pizan, C. de (1405). *Le Livre de la Cité des Dames (The book of the city of women)*. Paris.

Protze, K. (2009). *Hausen statt Wohnen. Von der Hartnäckigkeit gesellschaftlicher Wertvorstellungen in wechselnden Leitbildern – Vorschlag für einen Blickwechsel (Housing instead of living. The stubbornness of social values in changing models – suggestion for a change of perspective)*. Notizbuch der Kasseler Schule, Nr. 74. Kassel.

Rebe, S. (2002). *Aktuelle Frauenwohnprojekte in Deutschland – eine Dokumentation (Current women's housing projects in Germany – a documentation)*. Schriftenreihe Niedersächsisches Innenministerium. Hannover. [Online]. Available at: www.igt-arch.uni-hannover.de/fileadmin/pas/FORSCHUNG/Aktuelle_Frauenwohnprojekte.pdf [Accessed 13 February 2019].

Reeves, D. (2002). Mainstreaming gender inequality. An examination of the gender sensitivity of strategy planning in Great Britain. *Town Planning Review*, 73, 197–214.

Reichstein, F.M. (2001). *Das Beginenwesen in Deutschland. Studien und Katalog (The Beguinage in Germany. Studies and catalogue)*. Berlin: Verlag Dr. Köster.

Riss, S. (2017). *Frauengerechte Modellwohnprojekte der 1990er Jahre. Die versuchte Einflussnahme von Frauen als Impulsgeberinnen im österreichischen Wohnbau – Zusammenfassung (Model housing projects for women in the 1990s. The attempted influence of women as initiators in Austrian housing construction – abstract)*. Dissertation. Technische Universität Wien.

Roberts, M. (2013). Introduction. Concepts, themes and issues in a gender approach in planning. In: I. Sánchez de Madariaga and M. Roberts (Eds.), *Fair shared cities. The impact of gender planning in Europe*. Farnham: Ashgate, 1–18.

Roberts, M. (1991). *Living in a man-made world. Gender assumptions in modern housing design*. London: Routledge.

Rodenstein, M. (1994). *Wege zur nicht-sexistischen Stadt. Architektinnen und Planerinnen in den USA (Ways to a non-sexist city. Architects and planners in the USA)*. Freiburg i. Br.: Kore.

RTPI – Royal Town Planning Institute (2007). *Gender and spatial planning. Good practise note 7*. London.

Sailer, K. (2004). *Raum beißt nicht. Neue Perspektiven zur Sicherheit von Frauen im öffentlichen Raum (Space does not scare. New perspectives on safety of women in public space).* Beiträge zur Planungs- und Architektursoziologie, Bd. 2. Frankfurt/Main: Peter Lang.

Sainsbury, D. and Bergqvist, C. (2009). The promise and pitfalls of gender mainstreaming. *International Feminist Journal of Politics*, 11, 216–234.

Sapir, A. (2006). Globalization and the reform of European social models. *Journal of Common Market Studies*, 44(2), 369–390.

Scheller, A. (1995). *Frau – Macht – Raum. Geschlechtsspezifische Regionalisierung der Alltagswelt als Ausdruck von Machtstrukturen (Women – power – space. Gender-specific regionalisation of the everyday world as expression of power relations).* Zürich: Geographisches Institut der Universität Zürich.

Schmidt-Waldherr, H. (1999). Emanzipation durch Küchenreform? Einküchenhaus versus Küchenlabor (Emancipation through kitchen reform? One-kitchen building versus kitchen laboratory). *L'Homme. Zeitschrift für feministische Geschichtswissenschaft*, 1, 57–76.

Schnitter, B. (1984). Möglichkeiten einer Frauenarchitektur (Possibilities of women's architecture). *Schweizer Ingenieur und Architekt*, 50(84), 1011–1015.

Schönwandt, W.L. (2008). *Planning in crisis? Theoretical orientations for architecture and planning.* Hampshire: Ashgate.

Schönwandt, W.L. (2002). *Planung in der Krise? Theoretische Orientierungen für Architektur, Stadt- und Raumplanung.* Stuttgart: Kohlhammer.

Schröder, A. (2009). *Gender Mainstreaming als Instrument bedarfsgerechter Wohnraumversorgung. Prozesse, Strukturen und Projekte anhand ausgewählter Kommunen (Gender mainstreaming as instrument of housing provision as required. Processes, structures and projects on the basis of selected municipalities).* Hamburg: Kovac.

Schröder, A. and Zibell, B. (2004). *Auf den zweiten Blick. Städtebauliche Frauenprojekte im Vergleich (At second glance. Comparison of women's urbanistic projects).* Schriftenreihe "Beiträge zur Planungs- und Architektursoziologie", Bd. 1. Frankfurt: Peter Lang.

SIA – Schweizerischer Ingenieur- und Architektenverein (2015). *SIA-Merkblatt 2050 und Dokumentation 0246. Nachhaltige Raumentwicklung – Kommunale und regionale Planung (SIA – Swiss Association of Engineers and Architects. Leaflet 2050 and documentation 0246. Sustainable spatial development – Local and regional planning).* Zürich: SIA (www.sia.ch).

SIA – Schweizerischer Ingenieur- und Architektenverein (2005). *SIA 112/1. Nachhaltiges Bauen – Hochbau – Verständigungsnorm zu SIA 112. (SIA – Swiss Association of Engineers and Architects. Standard guide to SIA norm 112 – Sustainable construction – Structural engineering.)* Zürich: SIA (www.sia.ch).

Siemonsen, K. and Zauke, G. (1991). *Sicherheit im öffentlichen Raum. Städtebauliche und planerische Massnahmen zur Verminderung von Gewalt (Safety in public space. Urbanist and planning measures to reduce violence).* Zürich/Dortmund: ed. ebersbach im eFeF-Verlag.

Simons, W. (2001). *Cities of ladies. Beguine communities in the medieval low countries 1200–1565.* Philadelphia: University of Philadelphia Press.

Sitte, C. (1889). *Der Städtebau nach seinen künstlerischen Grundsätzen (Urbanism according to its artistic principles).* Basel: Birkhäuser.

SRL FG Frauen – Vereinigung für Stadt-, Regional- und Landesplanung Fachgruppe Frauen (1999). *Feministische Strategien für nachhaltige Stadt- und Regionalentwicklung – Debatten um Indikatoren in Deutschland. Ergebnisse eines Runden Tisches von Planerinnen in der Bundesrepublik Deutschland in Bonn, 27. November 1998 (Feminist strategies for sustainable urban and regional development – Debates on indicators in Germany. Results of a round table of female planners in the Federal Republic of Germany in Bonn, 27th of November, 1998)*. Berlin: SRL e.V.

Strüver, A. (2018). Performing spaces – gender relations, corpo-realities and space. In: C. Onnen and S. Rode-Breymann (Eds.), *Wiederherstellen – Unterbrechen – Verändern? Politiken der (Re-)Produktion (Restoring – interrupting – changing? Policies of (re-)production)*. L'AGENda Bd. 3. Opladen: Verlag Barbara Budrich, 79–91.

Stübben, J. (1890). *Der Städtebau (Urbanism)*. Braunschweig: Vieweg.

Studer, H. (2002). Mädchenräume – landschaftsplanerische Erfahrungen (Girls' spaces – experiences in landscape planning). In: C. Kramer (Ed.), *FREI-Räume und FREI-Zeiten: Raum-Nutzung und Zeit-Verwendung im Geschlechterverhältnis (Use of space and time in matters of gender relations)*, Schriften des Heidelberger Instituts für Interdisziplinäre Frauen- und Geschlechterforschung (HIFI) e.V. Baden-Baden: Nomos-Verl.-Ges.

Terlinden, U. (Ed.) (2010). Soziologie und Planung – Notizen zu einem zunehmend engen und ambivalenten Verhältnis (Sociology and planning – Notes on an increasingly close and ambivalent relationship). In: A. Harth and G. Scheller (Eds.), *Soziologie in der Stadt- und Landschaftsplanung: Analysen, Bedeutung und Perspektiven. (Sociology in urban and landscape planning. Analyses, significance and perspectives)*. Wiesbaden: VS Verlag für Sozialwissenschaften, 69–86.

Terlinden, U. (2006). Symbolische Herrschaft und Geschlechterkulturen. In: M. Rodenstein (Ed.), *Das räumliche Arrangement der Geschlechter: Kulturelle Differenzen und Konflikte (The spatial arrangement of genders. Cultural differences and conflicts)*. Berlin: trafo Verlag, 193–205.

Terlinden, U. (1990). *Gebrauchswirtschaft und Raumstruktur. Ein feministischer Ansatz in der soziologischen Stadtforschung (Practical economy and spatial structure. A feminist approach in sociological urban research)*. Stuttgart: Silberburg.

Terlinden, U. (1980). Baulich-räumliche HERRschaft. Bedingungen und Veränderungen (Structural-spatial dominion of men. Conditions and changes). *Beiträge zur feministischen Theorie und Praxis*, 4, 48–62.

Terlinden, U., Grieser, S. and Ross, B. (1999). *Wohnungspolitik in der alten Frauenbewegung. Die wohnungs- und siedlungspolitische Debatte in der alten deutschen Frauenbewegung und ihr Beitrag zu den Wohn- und Siedlungsformen in Kaiserreich und Weimarer Republik (Housing policy in the ancient women's movement. The Housing and Settlement Policy Debate in the ancient women's movement in Germany and its contribution to the typologies of housing and settlement in the Empire and Weimar Republic)*. Arbeitsberichte des Fachbereichs Stadtplanung, Landschaftsplanung Heft A 137, Universität Kassel.

Terlinden, U. and Oertzen, S. von (2006). *Die Wohnungsfrage ist Frauensache! Frauenbewegung und Wohnreform 1870 bis 1933 (The housing question is a matter for women! Women's movement and housing reform 1870 to 1933)*. Berlin: Dietrich Reimer Verlag.

Tilia (2007). *Berücksichtigung von Gender Mainstreaming bei der Gestaltung von Parkanlagen und Bewegungsangeboten für ältere Menschen im öffentlichen Raum (Taking gender mainstreaming into account in the design of parks and opportunities for movement for older people in public spaces)*. Study conducted on behalf of Co-ordination Office for Planning and Construction Geared to the Requirements of Daily Life and the Specific Needs of Women of the Executive Group for Construction and Technology. Unpublished, Vienna.

Triest, M. (1998). *Het besloten hof. Begijnen in de Zuidelijke Nederlanden (The private court. Beguines in the Southern Netherlands)*. 2nd edn. Antwerpen: Van Halewyck.

Tummers, L. (2013). Gendered perspectives on spatial planning and housing in the Netherlands. In: I. Sánchez de Madariaga and M. Roberts (Eds.), *Fair shared cities. The impact of gender planning in Europe*. Farnham: Ashgate, 107–129.

Uhlig, G. (1981). *Kollektivmodell "Einküchenhaus". Wohnreform und Architekturdebatte zwischen Frauenbewegung und Funktionalismus 1900–1933 (Collective model "one-kitchen building". Housing reform and architectural debate between women's movement and functionalism 1900–1933)*. Werkbund Archiv 6. Gießen: Anabas Verlag.

Ullmann, F. (2013). Choreography of life. Two pilot projects of social housing in Vienna. In: I. Sánchez de Madariaga and M. Roberts (Eds.), *Fair shared cities. The impact of gender planning in Europe*. Farnham: Ashgate, 297–324.

Unger, H. (2005). *Die Beginen. Eine Geschichte von Aufbruch und Unterdrückung der Frauen (The Beguines. A story of the departure and oppression of women)*. Freiburg i.Br. Basel Wien: Herder Verlag.

United Nations (1992). *Agenda 21*. [Online]. Available at: https://sustainabledevelopment.un.org/content/documents/Agenda21.pdf [Accessed 4 Aug. 2018].

Unwin, Raymond (1909). *Town planning in practice. An introduction to the art of designing cities and suburbs*. London: T. Fisher Unwin.

Urbact (2006). *Handbook for participation. Participation of inhabitants in integrated urban regeneration programmes as a key to improve social cohesion*. [Online]. Available at: www.eukn.eu/eukn-research/policy-labs/public-participation-in-the-development-process/other-publications/european-handbook-on-participation-urbact/ [Accessed 2 July 2018].

Vestbro, D.U. (1997). Collective housing in Scandinavia – how feminism revised a modernist experiment. *Journal of Architectural and Planning Research*, 14(4), 329–342.

Vestbro, D.U. and Horelli, L. (2012). Design for gender equality. The history of cohousing ideas and realities. *Built Environment*, 38(3), 315–335.

Wankiewicz, H. (2013). European regional development programmes for cities and regions. Driving forces for gender planning? In: I. Sánchez de Madariaga and M. Roberts (Eds.), *Fair shared cities. The impact of gender planning in Europe*. Farnham: Ashgate, 131–141.

Wastl-Walter, D. (2010). *Gender Geographien: Geschlecht und Raum als soziale Konstruktionen (Gender geography. Gender and space as social constructions)*. Sozialgeographie kompakt. Stuttgart: Franz Steiner Verlag.

West, C. and Zimmerman, D.H. (1987). Doing gender. *Gender & Society*, 1(2), 125–151.

Wetterer, A. (2004). Konstruktion von Geschlecht. Reproduktionsweisen der Zweigeschlechtlichkeit (Construction of gender. Reproduction of bisexuality). In: R. Becker and B. Kortendiek (Eds.), *Handbuch Frauen- und Geschlechterforschung: Theorie, Methoden, Empirie (Handbook women's and gender studies. Theory, methods and empiricism)*, Reihe "Geschlecht & Gesellschaft", Bd. 35, 1st edn. Wiesbaden: VS Verlag für Sozialwissenschaften. 122–131.

Wotha, B. (2013). Urban governance and gender-aware planning. In: I. Sánchez de Madariaga and M. Roberts (Eds.), *Fair shared cities. The impact of gender planning in Europe*. Farnham: Ashgate, 91–130.

Wotha, B. (2000). *Gender Planning und Verwaltungshandeln. Umsetzung von Genderbelangen in räumliche Planung – unter Berücksichtigung von Verwaltungsmodernisierung und neuerer Tendenzen im Planungsbereich (Gender planning and administrative action. Implementation of gender requirements in spatial planning – taking into account administration modernisation and new trends in the planning sector)*. Kieler Arbeitspapiere zur Landeskunde und Raumordnung, Kiel.

Young, I.M. (2005). Justice and the politics of difference. In: S. Fainstein and L. Servon (Eds.), *Gender and planning. A reader*. New Brunswick, NJ, London: Rutgers University Press, 86–103.

Zibell, B. (2019). Architektur und Raumplanung. Zur Herausforderung der geschlechtergerechten Gestaltung eines Wissens- und Handlungsfeldes (Architecture and spatial planning. The challenge of gender-fair design in a field of knowledge and action). In: B. Kortendiek, B. Riegraf and K. Sabisch (Eds.), *Handbuch Interdisziplinäre Geschlechterforschung (Handbook of interdisciplinary gender research)*, Reihe "Geschlecht und Gesellschaft", Bd. 65. Wiesbaden: Springer VS, 709–718.

Zibell, B. (2013). The model of the European city in the light of gender planning and sustainable development. In: I. Sánchez de Madariaga and M. Roberts (Eds.), *Fair shared cities. The impact of gender planning in Europe*. Farnham: Ashgate, 75–90.

Zibell, B. (1993). Der Brahmshof in Zürich. Ein Wohnmodell des Evangelischen Frauenbundes Zürich (The Brahmshof in Zurich. A housing model of the Evangelischer Frauenbund Zurich). In: *Neue Wohn- und Siedlungsformen. Impulse aus Frauensicht. (New typologies of housing and settlement. Impulses from a women's position)*. Dokumentation des Symposions des Beirats für frauenspezifische Belange bei der Senatsverwaltung für Bau- und Wohnungswesen vom 29./30. Januar 1993 in Berlin. Berlin: SenBauWohn, 20–39.

Zibell, B. (1983). *Frauen in Wohnumfeld und Nachbarschaft (Women in living environment and neighbourhood)*. Arbeitshefte des Instituts für Stadt- und Regionalplanung der TU Berlin, Nr. 26. Berlin: Universitätsverlag TU Berlin.

Zibell, B., Damyanovic, D. and Àlvarez, E. (2016). *On stage! Women in landscape_architecture and planning*. Berlin: Jovis.

Zibell, B. and Schröder, A. (2007). *Frauen mischen mit. Qualitätskriterien für die Stadt- und Bauleitplanung (Women get involved. Quality criteria for urban and land-use planning)*. Frankfurt/Main: Peter Lang.

References to the country tables

Finland

Böhme, K. (2002). *Nordic echoes of European spatial planning. Discursive integration in practice*. Stockholm: Nordregio.

Hirdman, Y. (2001). *Genus – om det stabilas föränderliga former (Gender – changing forms of the stable)*. Malmö: Liber.

Hirvonen-Kantola, S. and Mäntysalo, R. (2014). The recent development of the Finnish planning system. The city of Vantaa as an executor, fighter and independent actor. In: M. Reimer, P. Getimis and H.H. Blotevogel (Eds.), *Spatial planning systems and practices in Europe. A comparative perspective on continuity and changes*. New York: Routledge, 42–60.

Ministry of Social Affairs and Health (2017). *Government Action Plan for Gender Equality 2016–2019*. Publications of the Ministry of Social Affairs and Health, Helsinki.

Netherlands

ARL/DGR (Ed.) (2003). *Deutsch-Niederländisches Handbuch der Planungsbegriffe (German-Dutch manual of planning terms)*. Hannover/Den Haag: Akademie für Raumforschung und Landesplanung, ARL.

Bouwnetwerk (undated). *Wat is Bouwnetwerk?* [Online]. Available at: www.bouwnetwerk.net/watisbnw.php [Accessed 3 Aug. 2018].

Netherlands Institute for Multiparty Democracy and Instituut voor Publiek en Politiek (2008). *The Dutch Political System in a Nutshell*. The Hague/Amsterdam: Netherlands Institute for Multiparty Democracy/Instituut voor Publiek en Politiek.

TU Delft (undated). *DEWIS*. [Online]. Available at: www.tudelft.nl/over-tu-delft/strategie/strategiedocumenten-tu-delft/diversiteitsbeleid/dewis/ [Accessed 3 Aug. 2018].

VACpunt Wonen (undated). *VACpunt Wonen*. [Online]. Available at: www.vacpuntwonen.nl [Accessed 3 Aug. 2018].

Vrouwen Bouwen Wonen (undated). *Welkom*. [Online]. Available at: https://vrouwen-bouwenwonen.weebly.com [Accessed 3 Aug. 2018].

England

Burgess, G. (2008). Planning and the gender equality duty. Why does gender matter? *People, Place and Policy*, 2(3), 112–121.

Ministry of Housing, Communities and Local Government (2015). *Plain English guide to the planning system*. [Online]. Available at: www.gov.uk/government/publications/plain-english-guide-to-the-planning-system [Accessed 3 Aug. 2018].

The Equilibrium network (undated). *The Equilibrium network*. [Online]. Available at: https://equilibrium-network.com/ [Accessed 3 Aug. 2018].

The Scottish Parliament (undated). *Visit and learn*. [Online]. Available at: www.parliament.scot/visitandlearn/Education/18642.aspx [Accessed 3 Aug. 2018].

URBANISTAS (undated). *About us*. [Online]. Available at: www.urbanistas.org.uk/about-us [Accessed 3 Aug. 2018].

Women in planning (undated). *About us*. [Online]. Available at: https://womeninplanning.wordpress.com/about/about-us [Accessed 3 Aug. 2018].

Austria

Genderplattform (undated). *Genderplattform.at*. [Online]. Available at: www.genderplattform.at [Accessed 3 Aug. 2018].

Germany

Pahl-Weber, E. and Henckel, D. (Eds.) (2008). *The planning system and planning terms in Germany. A glossary*. Hannover: Akademie für Raumforschung und Landesplanung, ARL.

France

EspaceSuisse (2018). *Gesprächsunterlage zur Aussensicht auf die Schweizer Raumplanung. Planungssysteme und Herausforderungen in Deutschland, Frankreich, Luxemburg, den Niederlanden und Österreich. (Discussion paper on the external view of Swiss spatial planning. Planning systems and challenges in Germany, France, Luxembourg, the Netherlands and Austria)*. Bern.

Geppert, A. (2014). France, drifting away from the "regional economic" approach. In: M. Reimer, P. Getimis and H.H. Blotevogel (Eds.), *Spatial planning systems and practices in Europe. A comparative perspective on continuity and changes*. New York: Routledge, 109–126.

The Interministerial Agency for Spatial Planning and Competitiveness (DIACT) and the Directorate of Development Policies at the Ministry of Foreign Affairs (2006). *Spatial planning and sustainable development policy in France*. Paris: Ministère des Affaires étrangères.

Zotero (2016). *Genre et urbanisme (Gender and urbanism)*. [Online]. Available at: www.zotero.org/groups/509948/genre_et_urbanisme [Accessed 3 Aug. 2018].

Spain

Col·lectiu Punt 6 (undated). *Articles and books*. [Online]. Available at: www.punt6.org/en/articles-and-books/ [Accessed 3 Aug. 2018].

Sánchez de Madariaga, I. (2004). *Urbanismo con perspectiva de género (Urbanism with a gender perspective)*. Sevilla: Fondo Social Europeo – Junta de Andalucía.

Greece

Athens Social Atlas (2000). *Athens Social Atlas*. [Online]. Available at: www.athenssocialatlas.gr [Accessed 3 Aug. 2018].

Giannakourou, G. (undated). *Planning law in Greece. A brief overview*. Thessaloniki: University of Thessaly.

La Spina, A. and Sciortino, G. (1993). Common agenda, southern rules. European integration and environmental change in the Mediterranean states. In J.D. Liefferink, P.H. Lowe and A.P.J. Mol (Eds.), European Integration and Environmental Policy. London/New York: Belhaven Press, 217–236.

Serraos, K., Gianniris, E. and Zifou, M. (2005). *The Greek spatial and urban planning system in the European context*. Athens: National Technical University of Athens, Faculty of Architecture, Department of Urban and Regional Planning.

Switzerland

ARL/VLP-ASPAN (2008). *Deutsch-Schweizerisches Handbuch der Planungsbegriffe (German-Swiss handbook of planning terms)*. Hannover/Bern: ARL/VLP-ASPAN.

VLP-ASPAN (2012). *Spatial planning in Switzerland: A short introduction*. Bern: VLP-ASPAN.

Several EU countries

European Commission (1997). *The EU compendium of spatial planning systems and policies*. Regional Development Studies, 28.

3 Gender mainstreaming and spatial development

Contradictions and challenges

Lidewij Tummers, Sylvette Denèfle and Heidrun Wankiewicz

Introduction

The EU-driven equality policies expressed in the Treaty of Amsterdam have been used in several countries in the first decade of the twenty-first century to innovate planning and spatial development and to introduce "gender" to planning and development processes of urban and rural areas (Wankiewicz, 2012, 2016). Yet the question of how urbanism can contribute to gender equality had already been raised in the 1980s (see for example: Hayden, 1983; Matrix, 1984; Paravicini, 1988). A number of research projects investigated urban environments (substance) and spatial development (process), while parallel attempts to make spatial policies and planning strategies gender-sensitive have produced a large number of high-quality experiments.

Together, research and practice form a body of theoretical and professional knowledge of which this book and earlier compilations contain numerous examples (see for example: Niranjana, 2001; Denèfle, 2004; Fainstein and Servon, 2005; Jarvis, Kantor and Cloke, 2009). However, the implementation of gendered planning criteria remains predominantly at the level of local programmes and projects. Although such projects are generally celebrated as "best practices", their structural embedding in planning systems and impact on planning cultures is unclear. In other words, the experiences and lessons from research and fieldwork are under-used. As Greed postulated in 2005, "making the link between spatial-policy and 'social' gender considerations proved one of the most difficult aspects of Gender Mainstreaming" (Greed, 2005: 262). While gender mainstreaming (GM) in urbanism is regarded as creating good quality urban space, the question arises of whether GM does in effect contribute to emancipatory goals.

This chapter offers a critical reflection on the implementation of gender-aware urbanism in European cities and regions. Building on the evaluations of local projects (such as, for example, Schröder and Zibell, 2004), we look into the frictions in urbanism's response to gender mainstreaming and identify controversies and ambiguities that arise. In doing so, we aim to overcome the difficulties these controversies produce and to advance thinking about the structural embedding of GM and long-term effective spatial strategies.

In this chapter we first introduce the background for our analyses and different perspectives on the production of space; then we present the types of controversies our research has found, and finally we discuss key elements to overcome possible deadlocks and move towards innovating planning practice based on feminist principles.

Background

The authors of this chapter observe a renewed interest in gendered approaches to urbanism in Austria, France and the Netherlands. In France, the revised law on gender equality of 4 August 2014[1] obliges local authorities to implement gender mainstreaming policies in all sectors, which constitutes an important incentive for larger cities such as Bordeaux,[2] Lille, Lyon and Paris to (re-)introduce programmes around "women and the city". In the Netherlands, students and professionals express a new interest in equal representation and gender equality issues, organising debates, expositions, publications and networking activities, etc. In Austria, gender planning has entered mainstream urbanism and spatial planning only in the City of Vienna, which has a GM office within the urbanism department. However, some Austrian cities and regions use the "gender+" concept,[3] which combines gender with other forms of structural inequalities, such as age, migrant background or disability.

Gender-aware spatial development requires, as Burgess (2008: 112) underlines, "a more explicit consideration of gender … in both how planning is delivered and in its wider impact". Most gender-aware planning proposals and instruments are still useful for this new interest, but a critical assessment of past experiences is relevant. In this chapter we ask how gender mainstreaming knowledge can be made more effective and lead to the structural embedding of gender awareness in urbanism. Moreover, we caution against a simple copy-and-paste of existing gender mainstreaming instruments, for two groups of reasons:

- First, social conditions in Europe have evolved and in many places dramatically degraded as a consequence of the neo-liberal policies of the twenty-first century. The economic, social, environmental and political crises force us to fundamentally rethink the use and distribution of resources. Policies and strategies addressing the crises in an integrated way are generally labelled "sustainable". This contribution departs from the understanding that sustainability and (gender) equality are fully compatible and mutually dependent. This is all the more urgent because spatial planning is based on long processes for decision-making and consultation, and takes considerable time to come into effect.
- Second, technological development and – especially – the entrance into the digital age require all approaches to urbanism dating from the 1990s to be revised in terms of communication and automation. While, on the one hand, new technologies can be used for consultation, participation and activism[4] in planning, they can also aggravate disparities and lead to

non-transparent communication and decision-making. In the design and application of algorithms, etc., the underlying assumptions, for example about family life-styles and aspirations, largely remain invisible.

Evaluations of gender mainstreaming[5] strategies have shown that GM is not structurally embedded in urban design and planning practice. Singling out the gender aspect often meets with resistance because planners perceive their practice as "planning for all" and therefore "not making any distinctions", rather than taking positive action which prioritises one group over another. Nevertheless, user groups such as low-income households, senior or disabled persons, may receive occasional special attention as vulnerable "special needs" target groups. Several evaluators of national policies (Larsson, 2006; Burgess, 2008; Cortolezis, 2010; Lacey *et al.*, 2013) point to a lack of clarity in the conceptualisation of gender and the absence of knowledge about the gender dimensions of socio-spatial relations. In her ten-year evaluation of the toolbox gender and development Moser clarifies: "The simplification and perceived 'technification' of gender planning was a conscious decision in the highly hostile climate in which it was developed both to reach practitioners, and to provide operational tools they could implement" (Moser, 2014: 20).

While "fairness" to some extent constitutes the European way of thinking about the role of urban planning, in practice the roles and instruments of planning systems show considerable variety[6] (see Chapter 2). These different planning cultures form the background of our analyses of gender-aware and feminist projects and initiatives. We have also used these as a heuristic device in the development of our argument, analysing the terminology that reveals the value system underlying planning cultures.[7] The team further combines the perspectives of a geographer (HW), a sociologist (SD) and an engineer (LT), each with a different perspective on who creates the city. In order to include the wide range of substantive concerns and processes as well as different notions of professionalism, we use "urbanism" as a comprehensive term to cover practices, policies and research on governance and design of the environment, including both urban and rural areas. Feminist planning implies a conceptualisation of urbanism as a profession that aims to create fair, just and equally accessible urban environments. However, the city is not made through planning alone.

Perspectives on who creates the city

Internationally, since the introduction of "gender theory", feminist scholars have addressed urbanism because they expect it to have an influence on spatial development and hence to be capable of changing spatial conditions, especially of improving social and spatial justice. Thus, planners are ascribed an active role in shaping societies. Sandercock even defined urban planning as "an unfinished social project whose task is to manage our coexistence in the shared space of cities and neighborhoods in such a way as to enrich human life and to work for social, cultural and environmental justice" (Sandercock, 2004: 134). But who is implementing this "project"? Who can be considered a planner and are planners the main agents who "manage our co-existence in shared space"? The studies that geographers, sociologists and planners have published since the 1980s underline the necessary but often contradictory roles of political decision makers, specialised professionals and inhabitants.

In geography "women", "gender" and feminist perspectives started to be taken into account in the 1970s. The influential concept of time geography and the timeline in spatial patterns[8] (Hägerstrand, 1970; Ellergård and Karlsson, 2010) was criticised by feminists for using masculine models of time-space patterns (Rose, 1993). The so-called "spatial turn" from the 1990s introduced relational and social concepts of the process of "producing space" and the importance of place (Werlen, 1987; Massey, 1994; Soja, 2009) while applying normative categories like spatial justice, inequalities and regional disparities from geography in social and cultural science. From a sociologist's perspective, urban space is produced through its manifold uses. Gender norm(ative)s related to societal ideological choices (politics, religions, economics) construct the spatial forms. In sociological terms, urbanism is part of Western socio-political logic and as technical professional practice belongs to the strictly organised frameworks of public regulations, market priorities, experts and models that steer urbanism.

From a planning perspective, urbanism is based on situated and contextualised knowledge. Its concepts and ideas cannot be properly analysed without taking into account the physical qualities of space; the material, location and dimensions of places and the institutional framework of technical and financial regulations. Consequently, any urban design needs to be understood in the context of its planning culture.

In the complex reality of cities whose development is driven by economic and social motives, places are thus shaped by two juxtaposed realms of activity:

- on the one hand, "planning and development strategies" generated by planning professionals, who create proposals and design projects, whether following market logic or political priorities, e.g. balancing economic development and nature preservation;
- on the other hand, "placemaking" by inhabitants and user groups who appropriate spaces by specific forms of use or assigning meanings in order to create identity and landmarks.

We emphasise that to take into account social diversity and, especially, gendered practices, it is not enough to deepen understanding of diverse user needs or "placemaking". The "professional" production of space, analyses and theories also needs to be "gendered". As we will see in the examples below, most approaches aim to combine these user and planner perspectives. Moreover, approaches have to be assessed in terms of assumptions concerning the role of planning: does planning reflect power relationships in society *or* is it seen as transformative – each leads to a different line of advocacy planning.

Furthermore, it is important to differentiate between the actual production of space, and applied and academic spatial analyses and research. Professionals need to respond to concrete and direct needs, while academic research is more explorative and less bound to time restrictions. Professionals have a demand for pre-designed, feasible solutions, while researchers argue there is no blueprint for gender planning in "Western" planning cultures, since long-term objectives must be translated to very diverse local conditions. Professionals and scholars will at times use the same type of sources, but are likely to prioritise different sets of information. Although urbanism involves both academic and professional practices, which mutually feed each other, there is also a structural gap between the type of knowledge they require and produce. Therefore we see an indispensable need to connect theory (research) and practice (policies and plans) in order to dismantle the way planning is delivered, as well as to design "ideal models" and to experiment with egalitarian or "non-sexist cities" (Hayden, 1983).

To summarise, from our multi-disciplinary perspective three types of attitudes shape the permanent transformation of urban space: the attitude of urbanists (town planners), the attitudes of residents/users of all genders and the political priorities, based on economic options, cultural models, etc. These attitudes include gender stereotypes which are reproduced unconsciously (and often unnoticed) in urbanism and architecture. Assumptions are made, for example, about the use of cars, which turn out to be largely based on white-collar male time-space patterns or on standard layouts of single-family housing. When local or regional authorities take action to implement equality-oriented planning strategies, the voids between ideal, theory and practice become perceptible.

Types of controversies and the challenges they present

Across different planning systems, scholars evaluating GM have found that a combination of these controversies form an explanation of the fact that GM has not advanced structurally in planning. Larsson, for example, indicates that the notion of "equal opportunities" is not adequate for developing gender-aware planning based on an evaluation of Swedish equal opportunity policies in planning (Larsson, 2006). It is too much associated with the early stages of gender theory, identifying the "needs of women" and trying to patch the holes. Implicitly the terms of reference remain the same, suggesting that when women catch up with men, equality will be reached. Moreover, the implications of gender mainstreaming are not sufficiently clear. A study of Berlin arrives at a similar conclusion after a pilot programme by the Berlin Senate Department for Urban Development. In order to upscale the experience, it is not only necessary to provide a budget and knowledge: further professional innovation is needed to develop indicators for the quality of urban space and infrastructure (Berlin Senate Department for Urban Development, 2011c). Finally, Cortolezis notes that the goals of gender mainstreaming in Austrian planning are not sufficiently clear (Cortolezis, 2010).

In this section we present three types of challenges and controversies that have been discussed in the academic dialogue underpinning this book:

- First, as we will see, gender awareness often means the better servicing of gender-specific needs, without transforming gender roles. The impact of planning decisions/interventions on existing inequalities remains uncertain.
- Second, there is an inherent tension between the local, direct need for action and responses to discriminated and vulnerable groups, and the long-term goals connected to fundamental notions of equality and a radically different economic model.
- Third, gendered planning is often seen as a way to empower women or to reduce discrimination against women. However, theoretical concepts such as intersectionality are not operational in everyday planning practice.

Below, we discuss these types of controversies in more detail, looking at gender-aware planning practices, following Lombardo and Meier (2006: 161) in that "the awareness of inconsistencies can be a powerful tool for sharpening the formulation of gender policies".

Type 1: stereotypes vs gender-specific needs

One of the strategies to make urbanism more gender-sensitive is to make the differences amongst men and women explicit. Box 3.1 presents a successful example of how urban services could become more effective through gender awareness. For Karlskoga's (Sweden) public services, looking at the needs of women corrected their "blind spot" of user groups and consequently clearing the snow from much-used pedestrian areas reduced the number of accidents.

Box 3.1 Jamställ/Amställdsnöröjning: snow-removal for equality

The town of **Karlskoga**, 30,000 inhabitants, 250 km west of Stockholm, was the first to implement "gender-sensitive snow-clearing" in 2010. The idea emerged during a training session on gender for technical staff, funded by the Swedish Association of Local and Regional Authorities (SKL). Wondering what the gender aspects of snow-removal could be, they discovered there was a lot to be done. According to the statistics, more women than men use public transport and take children to school in the morning, walking or cycling. Often they ended up in hospital after falling on the frozen pavements. Karlskoga reversed the routes of the snow-cleaning machines, prioritising sidewalks and cycle-paths over car-lanes, and starting at kindergartens and schools, hospitals and public offices where the staff is predominantly female. According to the Head of Department, Bosse Björk, the measure also resulted in a reduction in traffic accidents, more than expected (Hivert, 2016).

The feminist movement of the twentieth century began by putting women's interests – previously unrecognised – on the agenda. The "diversity" approach, identifying user groups by additional sets of socio-demographic categories, such as age or employment situation, further leads to the better servicing of the diverse mobility and spatial needs of residents. The strength of this approach is that it enables the avoidance of homogeneous categories and the development of new perspectives which lead to innovations. However, without proper empirical evidence, or even when based on today's statistics, the approach holds the risk of stereotyping, for example seeing all men as 9–5 jobbers or all care-givers as women. Thus identifying homogeneous groups and gender roles without the perspective of change can become role-confirming, and lead to victimising or even stigmatising. Such a lack of understanding can result in missed opportunities, fake innovations and failed solutions. The risk of emphasising the interests of women, rather than the (power-)relations between different genders,[9] is illustrated in the project "High E-Quality City" (HQE) of Bordeaux Metropole (Box 3.2). The project includes mapping urban spaces of anxiety for both women and men, but re-affirms that "there seems to be equal access to performing citizenship, however for certain groups, notably women, this is not obvious" (C2D, 2014: 10). In the final report, "women" re-appear as a homogeneous category, suffering from discrimination. As in many other cases, women's perception of urban space is reported to be "different from men's", in other words, implicitly, the male perception is re-established as the norm. This is particularly the case in

Box 3.2 Bordeaux Ville Haute Qualité Egalitaire (Bordeaux High E-Quality City)

In 2011 a study into the use and accessibility of the Bordeaux agglomeration revealed great differences between men and women (Rasselet, 2011). The Regional Board for Sustainable Development (Conseil de Dévelopement Durable (C2D) de La Communauté urbaine de Bordeaux (CUB)), instigated the working group "Gender & City" to develop concrete solutions to secure "equal access to urban resources" for all (CUB, 2014: 3). The working group looked at best practices from elsewhere, then set a priority goal: to "gradually transform the professional planning culture" (CUB, 2014: 18). In order to achieve this, two types of action were taken: city walks leading to different ideas on public space; and gender training for planning professionals. Quoting the research report, translated by the authors: "the study revealed the vulnerability of certain women in public space and the great inequality of women and men in their accessibility to the city and to its resources" (Rasselet, 2011: 6). The report points out that "individual emancipation and self-realisation is not as open and accessible as pretended, especially for women" because of their everyday lives (part-time work, children and senior care, lower incomes and so on…) which "are much more fragmented than those of men, which are more linear and regular" (Rasselet, 2011: 6). The City of Bordeaux continues to work on gender issues in planning, recently (Bordeaux Métropole, 2017) looking, for example, at mobility strategies.

gender-aware approaches that address security in urban public space. The right to bodily integrity is not secured in public space at all times for women or men, but while men are statistically at higher risk in public space, women are perceived more as potential victims and women's fear of going out at night is emphasised (Roberts, 2008).

To what extent can servicing needs that are otherwise overlooked have a wider impact in creating better urban quality, as well as reducing inequalities? To draw a parallel with other excluding mechanisms: improving wheelchair access, or orientation for visually impaired people *does* significantly increase mobility and autonomy for people with disabilities. Does "care-friendly planning" increase opportunities, qualities and possibilities for people performing care tasks as well as for those who are care dependent in the same way? To date there has been no substantial monitoring of results for spatial planning, and gender mainstreaming (as a young field of applied research) is no exception. Monitoring and evaluating results is a complex issue, to which GM could contribute with methods and criteria developed so far (see also Chapter 6). The EuroFem network, for example, was created to explore the long-term implications of planning based on the "infrastructure for everyday life" paradigm. The network assembled numerous examples to issue recommendations supported by professionals and researchers (Horelli, Booth and Gilroy, 2000).

The first type of controversies in gender planning can thus be summarised as the balancing act between "making gender explicit" and "stereotyping". Gender theory introduced the notion of "intersectionality", looking at the combination of multiple discrimination factors such as sex, race, age, occupation or class, but the way this is operationalised in planning practice as "diversity" ignores the dynamics of the concept and thus risks determining user groups on the basis of their current identities.

Type 2: strategic and tactical objectives

Garber and Turner found that "female local activism … has tended to mirror women's domestic concerns" (Garber and Turner, 1995: 26). Planning interventions building on this activism can improve spatial conditions for many women in their present situation. However, if action research is not able to take a sufficiently critical stance, such measures may become an obstacle for emancipation in the long term. Molyneux (1984), followed by Gilroy and Booth (1999), identified this as the dilemma between tactical (short-term) and strategic (long-term) action. Demanding a change in planning paradigm to recognise the complexity of everyday life, they proposed a holistic approach.

One recent example to illustrate the ambivalent role of urbanism, caught between direct needs and long-term objectives, is the proposal presented by the City of Paris (symposium held 13 October 2015) concerning sport, public lighting and "human" security. The planning department adopted a set of "equality" principles to design sport facilities and public spaces.[10] In practice, this involves addressing questions such as: For whom is this space meant? What are the needs of the surrounding population and potential user groups? Such questions can be answered by survey or participation sessions, amongst others. But there remain fundamental questions, such as: Will this design procedure enhance the equality or emancipation of the citizens involved? Which is the preferred option for planning sport facilities in terms of gender mainstreaming: to create football fields and encourage girls to enter a male-dominated sport, or to replace the land-use for football with investments in female-dominated sport facilities? So far, the approaches including social diversity have been: institutionalising participative processes, holding a survey and focus group interviews, observations on place-making to assess user demands. Urban planners could contribute additional approaches, such as sketching the spatial implications of both options and presenting them at a public hearing. This does not solve the dilemma but makes it visible and part of a locally embedded GM strategy.

Taking direct action through pilot projects does not exclude awareness of the need to address planning cultures as part of a long-term strategy – as the above-mentioned HQE programme also observed. Without institutional transformation, gender-aware projects remain incidents and ad-hoc events. Consequently, gender mainstreaming is asked to realise projects and interventions which require funding and building licences, while transforming the same financial and legal system they depend on. For this reason, planning departments in a number of European cities have published handbooks that should enable urbanists and professionals to implement gender mainstreaming in planning practice (see Box 3.3). Some of the handbooks are connected with funding opportunities or are conditional for obtaining finance (for example various EU funds).

The impressive number of planning handbooks confirms the expertise on urban design that has been derived from the pilot projects. On the other hand, the recommendations and criteria are often of a generic nature to make the handbook applicable to all sorts of planning situations. They may lack measurable criteria to evaluate or compare design proposals. Moreover, digital or hardcopy handbooks can be ignored or misunderstood if they are not brought to the attention of the intended users. Therefore, professional dialogue has also been encouraged, in the form of workshops, seminars, planning audits and so on. To facilitate this dialogue, the "Gender Kompass Planung" (Gender Compass Planning) was designed in 2008 by Zibell and Schröder (undated), of the Leibniz University in Hannover, commissioned by the Gender Mainstreaming and Planning Departments of the southern German City of Freiburg (Box 3.4).

Box 3.3 Selected handbooks and manuals for gender planning

Brussels: Falu, A. (2018). *Espaces métropolitains égalitaires (Gender equal metropolitan spaces)*. Available in Spanish, English and French at: www.metropolis.org/fr/nouvelles/2018/07/03/3550.

Paris: Ville de Paris (2016). *Guide Référentielle. Genre & espace public (Reference guide for gender in public space)*. Available at: http://api-site-cdn.paris.fr/images/86068.

Prague: Women Public Space Prague (2016). *How to design a fair shared city?* Available at: www.wpsprague.com/fairsharedcity/.

Barcelona: Col·lectiu Punt 6 (2014). *Mujeres trabajando. Guia de reconozimiento urbano con perspectiva de genero (Women working. Urban assessment guide from a gender perspective)*. Available in several languages at: http://issuu.com/punt6/docs/mujerestrabajando.

Berlin: Berlin Senate Department for Urban Development (Ed.) (2011a). *Gender mainstreaming in urban development. Berlin handbook.* Berlin: Kulturbuch-Verlag. Available at: www.stadtentwicklung.berlin.de/soziale_stadt/gender_mainstreaming/download/gender_englisch.pdf.

Berlin Senate Department for Urban Development (Ed.) (2011b). *Gender mainstreaming in urban development. Gender criteria*. Available at: www.stadtentwick lung.berlin.de/soziale_stadt/gender_mainstreaming/download/kriterien_englisch.pdf.

Berlin Senate Department for Urban Development (Ed.) (2011c). *Gender in mainstreaming urban development. Berlin on the path towards becoming a metropolis worth living in for women and men.* Berlin. Available at: www.stadtentwicklung.berlin.de/soziale_stadt/gender_mainstreaming/download/gender_broschuere_englisch.pdf.

Hamburg: Behörde für Stadtentwicklung und Umwelt (Ed.) (undated). *Planungs empfehlungen der Fachfrauen (Planning recommendations of female experts)*. Hamburg: Freie und Hansestadt Hamburg. Available at: https://docplayer.org/27164536-Planungsempfehlungen-der-fachfrauen.html.

Salzburg: Zibell, B. with Dahms, N.-S. and Karácsony, M. (2006). *Bedarfsgerechte Raumplanung. Gender Practice und Kriterien in der Raumplanung. Endbericht. Langfassung (Requirement-oriented spatial planning. Gender practice and criteria in spatial planning)*. Materialien zur Raumplanung, Bd. 20. [Online] Salzburg: Land Salzburg. Available at: www.alpine-space.org/2000-2006/uploads/media/GENDERALP__ROSP_lang.pdf (German) www.gender-archland.uni-hannover.de/1212.html (English).

Vienna: Damyanovic, D., Reinwald, F. and Weikmann, A. (2013). *Gender mainstreaming in urban planning and urban development*. Werkstattbericht, Nr. 130A. Vienna: Urban Development Vienna, Municipal Department 18 (MA 18) – Urban Development and Planning. Available at: www.wien.gv.at/stadtentwick lung/studien/pdf/b008358.pdf.

England: RTPI – Royal Town Planning Institute (2007). *Gender and spatial planning. Good practise note 7*. London. http://rtpi.org.uk/media/1731629/gpn7-_gender_and_spatial_planning__2007_.pdf.

EuroFEM: Horelli, L., Booth, C. and Gilroy, R. (2000). *The EuroFEM toolkit for mobilizing women into local and regional development.* Helsinki: Helsinki University of Technology, Centre for Urban and Regional Studies. Available at: http://issuu.com/eva_alvarez/docs/12_eurofem_toolbox.

UN Habitat (2013). *Gender responsive urban planning and design*. HS Number 039/13E. Available at: https://unhabitat.org/books/gender-responsive-urban-planning-and-design/.

Box 3.4 Gender Kompass Planung (Gender Compass for Planning)

The Gender Compass is a short document based on gender planning criteria developed by the research project Frauen mischen mit: Qualitätskriterien für die Stadt- und Bauleitplanung (Women get in on the act: quality criteria for urban and development planning) (Zibell and Schröder, 2007). It addresses professionals working in urban development and building, both in the private sector and within the municipality. The Gender Compass itself is a compact flowchart, summarising the main criteria for gender-aware planning, based on the concept of the "city of short distances". It offers an explanation of gender mainstreaming and makes reference to the article of the German Building Act that requires gender aspects be taken into account. Design criteria are grouped in five themes: housing, accessibility, amenities, public space and safety; and at four territorial scales: city; district; neighbourhood and site. The generic criteria such as "densification around transport nodes" or "intergenerational functionality" aim to increase gender awareness in planning. The Gender Compass is meant to be used in workshops and brainstorming sessions, for specific guidance and planning proposals. Its compactness is both its advantage – it is a concise checklist to bring interdisciplinary professionals together – and its handicap: the indicators, which are very generic, make it difficult to identify gender aspects for non-experts: this works better as a take-away reminder from facilitated group sessions rather than as a "stand-alone" guide (Zibell and Schröder, undated).

The GM handbooks are based on the "everyday life" or "city of short distances" ("Stadt der kurzen Wege") concept. The concept is explained by Ciocoletto as the necessity to "construct a network of basic infrastructure such as outdoor spaces to meet in, education, health and everyday shopping, and prioritise the streets that assemble these services instead of thinking of pedestrian zones only in commercial terms" (Ciocoletto, 2014: 14; authors' translation). This concept presents a third type of controversy.

Type 3: the everyday life – short distances and global technology

Gender mainstreaming places the reconciliation of "work and home" at the heart of its strategies, based on the assumption that control over daily life is key to individual and household lifestyles.[11] Lange and Jurczyk found that the German discourse on the socio-economic aspects of balancing a job and family life "tend[s] to underestimate the structural conditions of establishing a professional career and building up a family, which are mostly structural barriers" (Lange and Jurczyk, 2006: 119). Enabling time-space patterns, not only for young families but also for seniors, singles, etc. depends on spatial conditions such as the proximity of everyday services and public transport. The Prague publication "How to Design a Fair Shared City?" for example illustrates this with "8 short stories based on equitable urban planning in everyday life" (Women Public Space Prague, 2016).

GM planning strategies, such as the design and participation consultancies of "Col·lectiu Punt 6"[12] in Catalunya (Box 3.3) or the Vienna municipal office for planning geared towards special needs (Box 3.3), also promote the "city of short distances". Italian cities like Bergamo and Bolzano, together with grassroots initiatives and academics, have taken up Hägerstrand's time geography concept and developed an urban model. The Italian time-policy re-organised governance, planning and mobility attempts to reconcile work schedules with the locations and opening hours of care infrastructures (Gelmini and Zambianchi, 2013; Mareggi, 2013). Gender mainstreaming and the everyday life concept gained importance in policymaking and governance with its adoption by the European Union in 1995 and the mandatory gender mainstreaming strategy for policies, EU programmes and projects (Wankiewicz, 2013). Through EU programmes such as Interreg within the GenderAlp! Project (Land Salzburg, 2007) or within the ESF3 Daily Routines Project (Tummers, 2012) it was introduced to other planning cultures. Box 3.5 introduces the concept of proximity in more rural (alpine) areas at regional scale-level in an example of cross-border regional collaboration. A similar example can be found in Nordregio – Dymén, Perio and Tepecik Diş (2015) extend the diversity of the everyday life to the city region and to strategic planning documents. The networks of basic infrastructure now need to be redesigned for the digital age, which for example includes more sophisticated ("smart") transport systems.

Box 3.5 City-Region of Proximity – GenderAlp! Inter-regional project

The Interreg project GenderAlp! aimed primarily to facilitate peer learning and knowledge transfer to local and regional actors in the alpine region. Its main instruments were collecting instructive practices and performing pilot measures elaborated by administrative staff and gender experts. Out of the transnational collaboration across very different planning cultures, administrative systems and traditions, a common understanding of key issues of gender-aware planning emerged. The cities and regions involved were able to translate "gender" into a workable concept for planning tasks and the development of selected areas: the connected and accessible city-region of short distances. By applying gender mainstreaming, a regional planning strategy for housing and work and a user-oriented connected city of proximity were created, amongst others:

- within the design process for a tramway extension in the City of Freiburg (Baden-Württemberg, DE);
- a planning approach for user-oriented and integrated urban development including care facilities, provision and recreation to engender a business park development (Lower Austria, AT);
- four pilot projects in gender budgeting and impact orientation in policymaking, for Munich (DE), Genova (IT), Salzburg and Upper Austria (AT).

Major success factors were: peer learning with gender-aware civil servants; training by doing and good practice exchange via documents and site visits (GenderAlp!, 2007).

The plea for decentralisation and human-scale, user-oriented urban design should thus not be understood as ignoring urban and regional scale-levels. Furthermore, Buckingham (2013) is one of those who suggest that there is convergence between the perspective of women and that of sustainability. If the compact city was born out of environmental concern with the increased mobility and urban sprawl produced by the functional city concept, proximity city is a criticism of the lack of recognition for the so-called "reproductive" activities of the care economy, and an alternative response to demographic and environmental challenges (Tummers and Zibell, 2012; Wankiewicz, 2012).

Discussion: gender-sensitive and feminist planning

In the above sections, we have outlined three types of controversies:

1 The balancing act between making gender explicit and stereotyping, which mostly leaves women more as part of the scene than part of the action
2 The dilemma between direct action and structural change, or the challenge for urbanism to design for a future that will be different from today, including gender roles
3 The comprehensive nature of everyday life, as expressed in the use of urban space, permeating all sectors and scales of planning practice, vs the sectorised and local manifestation (materialisation) of planning decisions

In addition, male planners resist gender planning approaches as positive discrimination. A major advantage of the "everyday" approaches based on time-space patterns is that they address everyday conditions for local people and visitors, no matter which nationality or position or citizenship, through urban design. In this sense, the concept of proximity is not new to urbanism, but is, for example, reminiscent of the idea of human-scale cities introduced by Jane Jacobs. In her famous essays Jane Jacobs vividly described how urban spaces are appropriated by residents and small commerce, and advocated design that enables life in public space (Jacobs, 1961). In our days, Blokland *et al.* (2015), amongst others, argue that expressing urban citizenship implies expressing difference, and caution against universal, all-inclusive ideals.

This makes it easier for urbanists to identify with GM approaches. At the same time, user-oriented planning parameters are used by planners, without utilising the label "gender", which introduces the risk of losing sight of the gender equality goals. Gender-sensitive planning and gender mainstreaming in spatial planning are then transformed into "inclusive" planning, and "gender" dynamics are no longer visible. In terms of improved planning quality and fairness in spatial development, in some planning cultures a user orientation and user-friendly planning processes and practices are relatively new and can be considered as progress towards spatial justice. Including gender mainstreaming tools and methods to address gender inequalities is then just a step away.

The pitfall of "proximity", however, is that everyday needs become confined to the small scale, which is no longer the reality for the majority of households. Gender issues or inclusive planning is then associated with the neighbourhood, rather than challenging larger economic dynamics or the regional infrastructure (Tummers, 2013). This may include ignoring wider (digital) dynamics and global communities that nowadays form part of the everyday life for most of the population. The action radius for daily life expands through new media, and social networks are globalised, especially for migrants. After research in urban renewal areas, for example, Listerborn (2007) developed the hypothesis that this may (partly) explain the absence of migrant women in local participation dialogues, as their everyday life extends to a different scale.

The dilemma between direct action and structural change draws attention to an important challenge for urbanism in general: on the one hand, to better articulate and service specific needs, based on vigilant activity patterns and social positions, such as domestic care tasks; and on the other hand, to design for a future when such tasks will be different – for example as a result of digitalisation. Bauhardt (2004) calls the concept of planning that aims to adjust power asymmetry between stakeholders equity planning. According to Fainstein and Fainstein, *equity* planning stands for "the new movement for urban change that calls for greater representation of disadvantaged groups in the governmental process and for the decentralisation of governmental policy making" (Fainstein and Fainstein, 1996: 271) and focuses on the substance of the programme, whereas *democratic* planning is mainly orientated towards the participatory process. In gender mainstreaming planning practices, process and substance go hand in hand, claiming both fair representation in decision-making and including care activities in briefings and design. Does this make a difference? As Cuthbert puts it, gender cannot "be considered an add-on to the designer's brief … but should be consciously welded into our perceptions of the world and how we live in it" (Cuthbert, 2011: 132). He goes on to refer to a paradigm shift comparable to that for "sustainable development".

Urbanism in itself cannot change gender relations, but understanding the inherent controversies of gender mainstreaming helps us to see the transformative potential of "gender" for spatial development. This calls for a radical reconsideration of fundamental aspects of urbanism: both in design criteria, to qualify a planning document/a design proposal/an outcome as "engendered", and in process criteria to identify the strategic vs pragmatic gender needs and empower different groups and genders to have a say in planning and decision-making. To achieve this, ensuring equal participation and fair representation in local and regional planning and development decisions and access to resources is essential. To paraphrase Fainstein and Servon (2005): women have become part of the scene, but are still not part of the action, such as that of urban decision-making.

Conclusions

This chapter has aimed to understand the inherent controversies of gender mainstreaming approaches in planning, in order to advance the transformative potential of GM. The EU countries have equality regulations and policies in place, but in practice their implementation meets with institutionalised resistance. The universal guidelines, agreed in international institutions such as the UN (Sustainable Development Goals 2017) and EU (Treaty of Amsterdam 1997), form the basis from which to create locally specific solutions for urban design and governance. GM is a first step, making gender relations explicit as part of planning processes, and encouraging institutions to become gender aware and gender balanced. We found that gender+ planning often seems to restore "good" user-oriented planning in places where this had disappeared into the background. Thus, gender mainstreaming in planning becomes a field of action that can correct major mistakes of, and respond to major changes in, society. If, however, the objective of gender mainstreaming in planning is a structural end to gender inequalities, this calls for feminist/transformative spatial planning. Urbanism for equity or urban justice in its turn needs to be gender-sensitive in order to be transformative. Our main conclusion concerns the implementation of feminist planning, which integrates principles of equality (social justice) and sustainability (environmental justice) in urban policies and spatial development. The complexity for feminist urbanism is that it requires the simultaneous deconstruction of both "gender" and "planning".

The abandonment of relatively straightforward causalities in favour of an understanding of complex realities brings about new challenges for spatial planning practice. The analyses of controversies indicate key fields for overcoming the deadlocks and developing feminist urbanism:

- Knowledge: Enhancing the understanding of "gender" and "intersectionality" amongst urbanist professionals in planning departments and private firms. In order to overcome stereotyping and making "men" the implicit norm and "women" the special needs group, "gender" and "intersectionality" need to produce operational categories for planning regulations, briefings and design criteria.
- Substance: To relate gender issues to and embed the everyday practices of households in all planning time horizons and scale dimensions. Whether planning is state managed or market led, the budget allocation of public departments and priorities of investors need to be evaluated against long-term equality criteria (gender auditing) overcoming short-term, single-purpose interests.
- Process: This requires the introduction of "gender governance" which enables citizenship and participation and leads planning systems (legislation, zoning categories, etc.) away from "patriarchal/expert-and-interest-group-led processes" to "civil society co-decisions and co-creation processes" (self-made urbanism/self-organisation/having a say in strategic planning).

Feminist urbanism can be seen as an interdisciplinary strategy not only to provide equal access to urban resources, but specifically to enhance the quality of decision-making, effectiveness in public expenditure and the production of durable, low-impact urban spaces, buildings and infrastructure. In other words: feminist urbanism addresses both the process and substance of planning, at project level as well as conceptually.

The cross-border exchange in the international working group set up by the Academy for Spatial Research and Planning (ARL) has facilitated the accumulation and growth of insights, leading to the above analyses. If language and cultural barriers continue to be crossed, it may be expected that the dilemmas and restrictions experienced by gender mainstreaming today will lead to the innovation of planning practice in the future.

Notes

1 La loi du 4 août 2014 pour l'égalité réelle entre les femmes et les hommes, Ministère des Affaires Sociales, de la Santé et des Droits des Femmes.
2 www.bordeauxmetropole.fr/sites/default/files/c2d/fichiers/genreville_web.pdf [Accessed 20 Sept. 2013].
3 This term (Gender+) was introduced by Verloo for the FP6 QUING project (2006–2011). Gender+ takes into account that gender is always overlaid (intersects) with other characteristics (such as age, origin, education and occupation, religion, etc.). Quoting the Madrid declaration: "With the use of the concept of gender+, we signal that we work with gender as intersected with other structural inequalities" (QUING Project, 2011: 1).
4 For example OSM(=Open Street Map)-based applications and special tools like the "SafetiPinApp" use tracking and tracing to re-appropriate space and make it safe and accessible for all people (www.safetipin.com) [Accessed 28 July 2018].
5 For an explanation of "gender mainstreaming" see Chapter 1.
6 See also: Reimer, Getimis and Blotevogel (2014); Ryser and Franchini (2015). This book, as product of a European IWG, sets the boundaries at European planning cultures to further understanding of a complex reality. For the authors of this chapter, this means focusing on a context that we are familiar with, without excluding a global perspective or assuming that the European perspective is the only or key reality.
7 For example, although roughly indicating a similar realm, "Städtebau" (Austria/Germany) does not involve the same institutions as "urbanisme" (France) and, although covering similar activities, the French "Aménagement du territoire" translates as "territorial organisation and development", while the Dutch "Ruimtelijke ordening" emphasises regulation and ordering.
8 This conceptualisation of space not only introduces daily and weekly time paths into spatial thinking, but also different constraints for people according to their sex/gendered roles and tasks, e.g. domestic chores and care work, as well as constraints caused by the infrastructures of the everyday life opening hours of services (Hägerstrand, 1970).
9 Whereas "women" (or rather their activities) are often the blind spot in spatial plans, "gender" in the context of the spatial political economy does not concern "women" per se, but rather the organisation of industrial space and regional development based on a gendered division of labour.
10 Guide référentiel, Genre & espace public: http://api-site-cdn.paris.fr/images/86068 [Accessed 15 Nov. 2018].

11 https://eige.europa.eu/gender-mainstreaming [Accessed 30 July 2018].
12 Punt 6 takes its name from Article 2, Clause 6 of the Urban Renewal Law 2004 (F for female also being the 6th letter of the Latin alphabet).

References

Bauhardt, C. (Ed.) (2004). *Räume der Emanzipation (Places of emancipation)*. Wiesbaden: VS Verlag für Sozialwissenschaften.

Behörde für Stadtentwicklung und Umwelt (Ed.) (undated). *Planungsempfehlungen der Fachfrauen (Planning recommendations of female experts)*. Hamburg: Freie und Hansestadt Hamburg. Available at: https://docplayer.org/27164536-Planungsempfehlungen-der-fachfrauen.html.

Berlin Senate Department for Urban Development (Ed.) (2011a). *Gender mainstreaming in urban development. Berlin handbook*. Berlin: Kulturbuch-Verlag. Available at: www.stadt entwicklung.berlin.de/soziale_stadt/gender_mainstreaming/download/gender_englisch.pdf.

Berlin Senate Department for Urban Development (Ed.) (2011b). *Gender mainstreaming in urban development. Gender criteria*. Available at: www.stadtentwicklung.berlin.de/soziale_stadt/gender_mainstreaming/download/kriterien_englisch.pdf.

Berlin Senate Department for Urban Development (Ed.) (2011c). *Gender in mainstreaming urban development. Berlin on the path towards becoming a metropolis worth living in for women and men*. Berlin. Available at: www.stadtentwicklung.berlin.de/soziale_stadt/gender_mainstreaming/download/gender_broschuere_englisch.pdf.

Blokland, T., Hentschel, C., Holm, A., Lebuhn, H. and Margalit, T. (2015). Urban citizenship and right to the city. The fragmentation of claims. *International Journal of Urban and Regional Research*, 39(4), 655–666.

Bordeaux Métropole (2017). Semaine europeenne de la Mobilite. La question de la mobilite des femmes a l'affiche des rendez-vous de Bordeaux Métropole (European Mobility Week. The question of women's mobility on the dashboard of Bordeaux Métropole rendez-vous). Available at: www.bordeaux-metropole.fr/Espace-presse/Semaine-europeenne-de-la-Mobilite-la-question-de-la-mobilite-des-femmes-a-l-affiche-des-rendez-vous-de-Bordeaux-Metropole [Accessed 15 Nov. 2018].

Buckingham, S. (2013). Gender, sustainability and the urban environment. In: I. Sánchez de Madariaga and M. Roberts (Eds.), *Fair shared cities. The impact of gender planning in Europe*. Farnham: Ashgate, 21–32.

Burgess, G. (2008). Planning and the gender equality duty – why does gender matter? *People, Place and Policy*, 2(3), 112–121.

C2D – Conseil de Developpement Durable (2014). *Ville HQÉ – Haute Qualité Égalitaire, méthode de diagnostic territorial participatif et genre (The High E-Quality City. A method for participative and gendered spatial diagnosing)*. [Online]. Available at: http://c2d.bordeaux-metropole.fr/Travaux-passes/Travaux-a-l-initiative-du-C2D/Ville-HQE-Haute-Qualite-Egalitaire-2014 [Accessed 30 July 2018].

Ciocoletto, A. (2014). *Espacios para la Vida Cotidiana. Auditoria de Calidad Urbana con Perspectiva de Género (Public space for everyday life. Urban quality audit with gender perspective)*. [Online]. Barcelona: Col·lectiu Punt 6, 2014. Available at: www.issuu.com/punt6/docs/espaciosparalavidacotidiana [Accessed 27 July 2018].

Col·lectiu Punt 6 (2014). *Mujeres trabajando. Guia de reconozimiento urbano con perspectiva de genero (Women working. Urban assessment guide from a gender perspective)*.

[Online]. Available in several languages at: www.issuu.com/punt6/docs/mujerestraba jando [Accessed 27 July 2018].

Cortolezis, H. (2010). Gleichstellung als Ziel. Steht Regionalentwicklung drauf, muss Gender Mainstreaming rein (Equality as objective. If it is called regional development, it needs to hold gender mainstreaming). *RAUM*, 77, 26–29.

CUB – Communauté urbaine de Bordeaux (2014). *Reportage "Ville HQE"*. [Online]. Available at: www.lacub.fr/c2d/ [Accessed 27 Nov. 2014].

Cuthbert, A. (2011). *Understanding cities. Method in urban design*. London and New York: Routledge.

Damyanovic, D., Reinwald, F. and Weikmann, A. (2013). *Gender mainstreaming in urban planning and urban development*. Werkstattbericht, Nr. 130A. [Online]. Vienna: Urban Development Vienna, Municipal Department 18 (MA 18) – Urban Development and Planning. Available at: www.wien.gv.at/stadtentwicklung/studien/pdf/b008358.pdf [Accessed 15 Nov. 2018].

Denèfle, S. (Ed.) (2004). *Femmes et Villes (Women and cities). Collection Villes et Territoires*, N°8. Tours: Presses Universitaires François Rabelais.

Dymén, C., Perio, L. and Tepecik Diş, A. (2015). *City-region planning for everyday life. Experiences from four Nordic city-regions*. Stockholm: Nordregio.

Ellergård K. and Karlsson, K. (Eds.) (2010). *Proceedings of the Sustaining Everyday Life Conference. April 22–24 2009, Campus Norrköping, Sweden*. [Online] Linköping Electronic Conference Proceedings, 38. Linköping: Linköping University Electronic Press. Available at: www.ep.liu.se/ecp/038/ecp09038.pdf [Accessed 20 Sept. 2013].

Fainstein, S.S. and Fainstein, N. (1996). City planning and political values. An updated view. In: S. Campbell and S.S. Fainstein (Eds.), *Readings in planning theory*. Cambridge, MA: Blackwell, 265–287.

Fainstein, S.S. and Servon, L. (Eds.) (2005). *Gender and planning. A reader*. New Brunswick, NJ, London: Rutgers University Press.

Falu, A. (2018). *Espaces métropolitains égalitaires (Gender equal metropolitan spaces)*. [Online]. Available in Spanish, English and French at: www.metropolis.org/fr/nouvelles/2018/07/03/3550 [Accessed 30 July 2018].

Garber, J.A. and Turner, R.S. (Eds.) (1995). Gender in urban research. *Urban Affairs Annual Review*, 42. London, Thousand Oaks, CA, New Delhi: Sage.

Gelmini, F. and Zambianchi, M. (2013). A history, concepts and practice of time policies and time planning. The Bergamo case. In: I. Sánchez de Madariaga and M. Roberts (Eds.), *Fair shared cities. The impact of gender planning in Europe*. Farnham: Ashgate, 265–272.

GenderAlp! (2007). *Herzlich willkommen zu GenderAlp! Austria (Welcome to GenderAlp! Austria)* [Online]. Available at: www.genderalp.at/ [Accessed 11 Mar. 2018].

Gilroy, R. and Booth, C. (1999). Building an infrastructure for everyday lives. *European Planning Studies*, 7(3), 307–324.

Greed, C. (2005). An investigation of the effectiveness of gender mainstreaming as a means of integrating the needs of women and men into spatial planning in the United Kingdom. *Progress in Planning*, 64(4), 239–321.

Hägerstrand, T. (1970). What about people in regional science? *Papers of the Regional Science Association*, 24(1), 6–21.

Hamburg Behörde für Stadtentwicklung und Umwelt (Department of Urban Development and Environment) (Ed.) (undated). *Planungsempfehlungen der Fachfrauen (Planning recommendations of female experts)*. [Online] Hamburg: Freie und Hansestadt Hamburg. Available at: https://docplayer.org/27164536-Planungsempfehlungen-der-fachfrauen.html [Accessed 11 Nov. 2018].

Hayden, D. (1983). The grand domestic revolution. A history of feminist designs for American homes, neighborhoods, and cities. *The American Historical Review*, 88(4), 1080–1081.

Hivert, A.-F. (2016). En Suède, des élus dénoncent un déneigement sexiste (In Sweden, local parliament rejects sexism in snow-removal). *Le Monde*, 2 December 2016, also Videolink: hOps://vimeo.com/77692813).

Horelli, L., Booth, C. and Gilroy, R. (2000). *The EuroFEM toolkit for mobilizing women into local and regional development.* [Online] Helsinki: Helsinki University of Technology, Centre for Urban and Regional Studies. Available at: http://issuu.com/eva_alvarez/docs/12_eurofem_toolbox [Accessed 30 July 2018].

Jacobs, J. (1961). *The death and life of great American cities.* New York: Random House.

Jarvis, H., Kantor, P. and Cloke, J. (2009). *Cities and gender.* Abingdon, Oxon: Routledge.

Lacey, A., Reeves, D.E., Tankel, Y. and Miller, R. (2013). From gender mainstreaming to intersectionality. Advances in achieving inclusive and safe cities. In: C. Whitzman, C. Legacy, C. Andrew, F. Klodawsky, M. Shaw and K. Viswanath (Eds.), *Building inclusive cities. Women's safety and the right to the city.* London: Routledge, 143–161.

Land Salzburg (Ed.) (2007). *GenderAlp! Location factor human.* Salzburg: City of Salzburg.

Lange, A. and Jurczyk, K. (2006). Family, work and the welfare state under conditions of blurring boundaries in Germany. In: G. Rossi (Ed.), *Reconciling family and work. New challenges for social policies in Europe.* Mailand: FrancoAngeli, 119–140.

Larsson, A. (2006). From equal opportunities to gender awareness in strategic spatial planning. *Town Planning Review*, 77(5), 507–530.

Listerborn, C. (2007). Who speaks? And who listens? The relationship between planners and women's participation in local planning in a multi-cultural urban environment. *GeoJournal*, 70(1), 61–74.

Lombardo, E. and Meier, P. (2006). Gender mainstreaming in the EU. Incorporating a feminist reading? *European Journal of Women's Studies*, 13, 151–166.

Mareggi, M. (2013). Planning times of the city. An overview on urban time policies. In: *REAL CORP 2013–18th International Conference on Urban Planning, Regional Development and Information Society.* [Online] 701–710. Available at: www.corp.at/archive/CORP2013_68.pdf [Accessed 20 Oct. 2015].

Massey, D. (1994). *Space, place and gender.* Minneapolis: University of Minnesota Press.

Matrix (1984). *Making space. Women and the man-made environment.* London and Sydney: Pluto Press.

Molyneux, M. (1984). Mobilisation without emancipation? Women's interests, state and revolution in Nicaragua. *Critical Social Policy*, 4(10), 59–71.

Moser, C. (2014). *Gender planning and development. Revisiting, deconstructing and reflecting.* [Online] DPU60 Working Paper Series: Reflections. London: The Bartlett Development Planning Unit. Available at: www.bartlett.ucl.ac.uk/dpu [Accessed 7 Dec. 2015].

Niranjana, S. (2001). *Gender and space.* Delhi: Sage.

Paravicini, U. (1988). *Femmes et architecture domestique. Une Histoire matérielle de l'habitat (Women and domestic architecture. A material history of the habitat of the home).* Dissertation. École polytechnique fédérale de Lausanne.

QUING Project (2011). *Madrid declaration on advancing gender+ training in theory and practice.* Author: Verloo, M. [Online]. Available at: www.quing.eu/files/madrid_declaration.pdf [Accessed 15 Mar. 2013].

Rasselet, C. (Ed.) (2011). *L'usage de la ville par les femmes (The use of the city by women).* Bordeaux 3 and le Laboratoire de recherche Adess-CNRS (Aménagement, Développement, Environnement, Santé, Société). [Online]. Available at: www.aurba.org/Etudes/

Themes/Populations-et-modes-de-vie/L-usage-de-la-ville-par-le-genre-lesfemmes [Accessed 27 Nov. 2014].

Reimer, M., Getimis, P. and Blotevogel, H.H. (Eds.) (2014). *Spatial planning systems and practices in Europe. A comparative perspective on continuity and changes*. New York: Routledge.

Roberts, M. (2008). *Gender and the night-time city. Changing paradigms and their implications for planning*. Presented at the Urban Futures Madrid, 1–12.

Rose, G. (1993). *Feminism and geography. The limits of geographical knowledge*. Minneapolis: University of Minnesota Press.

RTPI – Royal Town Planning Institute (2007). *Gender and spatial planning. Good practise note 7*. London. Available at: http://rtpi.org.uk/media/1731629/gpn7-_gender_and_spatial_planning__2007_.pdf [Accessed 15 Nov. 2018].

Ryser, J. and Franchini, T. (Eds.) (2015). *International manual of planning practice*. 2015 edition. The Hague: International Society of City and Regional Planners, ISOCARP.

Sandercock, L. (2004). Towards a planning imagination for the 21st century. *Journal of the American Planning Association*, 70, 133–141.

Schröder, A. and Zibell, B. (2004). *Auf den zweiten Blick. Städtebauliche Frauenprojekte im Vergleich (At second glance. Comparison of women's urbanistic projects)*. Schriftenreihe "Beiträge zur Planungs- und Architektursoziologie", Bd. 1. Frankfurt am Main: Peter Lang.

Soja, E.W. (2009). The city and spatial justice. *Journal on Spatial Justice and Planning*, 1, 31–38.

Tummers, L. (2013). Urbanism of proximity. Gender-expertise or shortsighted strategy? Re-introducing gender impact assessments in spatial planning. *Tria Territorio della Ricerca su Insediamenti e Ambiente. Rivista Internazionale di Cultura Urbanistica*, 10(1), 213–218.

Tummers, L. (2012). The everyday as spatial development potential. In: C.-H. Hauptmeyer (Ed.), *Neue Chancen für Kommune und Region (New opportunities for cities and regions)*. Stadt und Region als Handlungsfeld, 11. Frankfurt am Main: Peter Lang, 69–81.

Tummers, L. and Zibell, B. (2012). What can planners do for the connected city? A gendered reading of the New Charter of Athens. *Built Environment*, 38(4), 524–539.

UN Habitat (2013). *Gender responsive urban planning and design*. HS Number 039/13E. [Online]. Available at: https://unhabitat.org/books/gender-responsive-urban-planning-and-design/ [Accessed 30 July 2018].

Ville de Paris (2016). *Guide Référentielle. Genre & espace public (Reference guide for gender in public space)*. [Online]. Available at: http://api-site-cdn.paris.fr/images/86068 [Accessed 30 July 2018].

Wankiewicz, H. (2016). *Gender Planning – Gender Mainstreaming in der räumlichen Planung. Top-down- und Bottom-up-Strategien als Bausteine zu einer nutzerInnen- und gleichstellungsorientierten feministischen Raumplanung (Gender planning – gender mainstreaming in spatial planning. Top-down and bottom-up strategies as part of a user and gender equality oriented feminist spatial planning)*. Dissertation. Paris Lodron Universität Salzburg.

Wankiewicz, H. (2013). European regional development programmes for cities and regions. Driving forces for gender planning? In: I. Sánchez de Madariaga and M. Roberts (Eds.), *Fair shared cities. The impact of gender planning*. Farnham: Ashgate, 131–155.

Wankiewicz, H. (2012). City region of short distance for ALL? Planning the "everyday" for a diversity and mixity of users in functional areas. In: *REAL CORP 2012–17th*

International Conference on Urban Planning, Regional Development and Information Society. [Online] 109–119. Available at: www.corp.at/archive/CORP2012_198.pdf [Accessed 21 Feb. 2017].

Werlen, B. (1987). *Sozialgeographie (Social geography)*. Bern: Hauptverlag.

Women Public Space Prague (2016). *How to design a fair shared city?* [Online]. Available at: www.wpsprague.com/fairsharedcity/ [Accessed 30 July 2018].

Zibell, B. with Dahms, N.-S. and Karácsony, M. (2006). *Bedarfsgerechte Raumplanung. Gender Practice und Kriterien in der Raumplanung. Endbericht. Langfassung (Requirement-oriented spatial planning. Gender practice and criteria in spatial planning)*. Materialien zur Raumplanung, Bd. 20. [Online] Salzburg: Land Salzburg. Available at: www.alpine-space.org/2000-2006/uploads/media/GENDERALP__ROSP_lang.pdf (German) [Accessed 9 May 2018], www.gender-archland.uni-hannover.de/1212.html (English) [Accessed 19 Nov. 2018].

Zibell, B. and Schröder, A. (2007). *Frauen mischen mit. Qualitätskriterien für die Stadt- und Bauleitplanung (Women get in on the act. Quality criteria for urban and development planning)*. Frankfurt/Main: Peter Lang.

Zibell, B. and Schröder, A. (undated). *Gender Kompass Planung. Handreichung (Gender compass planning. Handout)*. [Online] Freiburg: Stadt Freiburg. Available at: www.gender-archland.uni-hannover.de/1212.html [Accessed 15 Nov. 2018].

4 Gender sensitivity in urban development concepts

The example of two case studies from London and Vienna

Florian Reinwald, Marion Roberts and Eva Kail

Aims and content

The intention of this chapter is to examine how gender-sensitive policies and intentions are visible at the strategic city-wide level and in major development sites. The focus will be on the different approaches adopted in the UK and Austria. Since planning is multi-scalar, the chapter will move from a brief résumé at national and regional scales to a deeper consideration of two case studies.

The two central research questions for this chapter are: How was gender mainstreaming implemented at different spatial development levels such as state level in Austria and the UK and city level in Vienna and London? How was gender mainstreaming implemented and which criteria of (implicit) gender-sensitive planning are visible in urban master plans comparing Barking Riverside (London) and Aspern Seestadt (Vienna)?

The sites that have been chosen for comparison are both brownfield urban extensions. The planning intention in both cases is to produce a place with coherence and an identity that responds to the demands of everyday life in the twenty-first century. Both have required a significant investment in new infrastructure and while no site is a blank sheet, in both cases there was no significant historic development on the site to be retained.

Introduction – overview of urban development and gender sensitivity in the UK and Austria

Since the UK and Austria have differing political systems and organs of government, they will be discussed separately to amplify the information set out in Chapter 2 of this volume. A key difference lies in the federal structure of Austria with its nine states, which contrasts with the UK's union of three countries and Northern Ireland. For the sake of brevity this section focuses on England and Wales.

Overview of urban development and gender sensitivity in the UK

Planning legislation in England and Wales will "sit" under European legislation until the UK leaves the European Union in 2019. The UK fully adopted the

Treaty of Amsterdam, first through a Gender Equality Duty in 2007 and then through the Equality Act 2010, implemented in 2011. This Act covers many aspects of difference, with the "protected characteristics" being defined as gender, race, disability, age, sexual orientation, gender reassignment, religion or belief. The Act requires gender mainstreaming to apply to all aspects of public policy.

Planning in the UK is a quasi-legal process. A national strategic document, the National Planning Policy Framework (NPPF) guides locally determined plans and policies (CLG, 2012). Local plans are made by individual planning authorities but have to be approved by a Planning Inspector acting for the Secretary of State in central government. The NPPF makes no explicit reference to gender. It does, however, support many principles of gender-sensitive development, such as the need to provide a variety of housing types to support local needs, to provide for a mix of uses, to promote accessibility between residential uses and the local commercial, social and cultural infrastructure and to encourage a modal switch in favour of walking and cycling.

Local spatial plans are mainly written documents and incorporate a degree of flexibility in their implementation. They consist of a number of separate documents which combine to form the Local Development Framework. This includes a Statement of Community Involvement, which sets out the bodies the local authority intends to consult and an Equalities Impact Assessment, which is now being carried out in conjunction with other impact assessments before the plan is approved and adopted. The only area in England and Wales which has a regional planning authority as well as local planning authorities is London. Greater London is governed by an Assembly and a Mayor with 32 boroughs plus the City of London as the tier below them. Boroughs can appoint their own Design Review Panels to assess major master plans and development proposals, but there is no requirement for gender sensitivity to be addressed.

Overview of urban development and gender sensitivity in Austria

The first impulse for the implementation of gender mainstreaming in politics and administration was derived from the Treaty of Amsterdam 1997/1999. In 2000 the Ministerial Council decided on the implementation of gender mainstreaming in administration and an inter-ministerial working group on gender mainstreaming (IMAG GMB) was set up. Between 2000 and 2007, all the federal states included gender mainstreaming in their spheres of action.

In accordance with the decentralised planning system in Austria, top-down gender mainstreaming has to be implemented by each federal state, local authority and city administration, and processed into gendered urban planning and development. Austria has neither national planning legislation nor national planning competences besides the provision of the main technical infrastructure (Art. 15 B-VG). Only a coordination board, the "Austrian Conference on Spatial Planning" (which is comparable to EUREK) formulates general guidelines – the

"Austrian Spatial Development Concept" (ÖROK, 2011). Gender mainstreaming is not mentioned here. In the chapter "Social Diversity and Solidarity" it is recommended that implicit gender+ related topics be taken into account with the approach "Good design/planning for all".

More relevant strategic planning principles are formulated at federal state level. The federal state is the supervising authority responsible for local development plans (Art. 118 B-VG). Mayors and municipal councils in Austria are responsible for land-use and zoning plans as well as building permits, and also for development concepts and the design of the planning process as a whole concerning their community. Zoning plans have to be approved by federal planning departments.

Therefore, the various federal states in Austria, the cities and municipalities have developed different instruments and tools to implement gender mainstreaming in urban planning and development – or not. An analysis of the urban development concepts of eight of the federal capitals and two other cities[1] has shown that only six out of ten use gender-sensitive language. Only three of these urban development concepts explicitly use the term gender mainstreaming.

Differences and similarities

Both of the planning systems in Austria and the UK are "top-down". Planning principles have to be specified and "broken down" into ever more specific criteria at each planning level. There is a necessity, therefore, to modify the criteria for every detailed step in the planning process. Gender planning has the aim to link different levels, to interweave implementation, experience and theoretical approaches as well as the perspectives from "top-down" and "bottom-up". Not only are statutory links needed, but it is important not to neglect the "bottom-up". In addition, initiatives and individuals need to be nourished and empowered (Sánchez de Madariaga, 2013). Both the Austrian and UK planning systems lack a consistent "roll-out" of explicit gender mainstreaming at all spatial scales in all sectoral aspects and levels of governance. In Austria, the degree of roll-out differs considerably between the federal states. A gender perspective is missing from the Austrian Spatial Concept, but the relevance of this concept for practice is limited due to the federal structure in spatial development in Austria. In the UK, this deficit is more serious because although gender mainstreaming is prescribed in national law, it has not been explicitly translated into the NPPF. In this way the "euphoria" associated with the formal adoption of the Treaty of Amsterdam by EU member states has been dissipated. (A shining exception is the impact of European Regional Development Policy and Funds, with their specific requirements for gender mainstreaming.)

Notwithstanding these caveats, implicit gender+-sensitive planning concepts have been incorporated into the mainstream of planning policies in both Austria and the UK. The next section considers how gender-sensitive concepts are operationalised in the development plans for both capital cities.

Gender sensitivity in overall strategic urban development concepts in London and Vienna

The next level of comparison involves the major planning instruments on a city level. The urban development concepts for the London Plan and the STEP 2025 Urban Development Plan for Vienna are discussed and compared.

Overview of gender sensitivity in urban development in London – the London Plan

The first iteration of the London Plan began in 2000 with the reinstatement of the Greater London Authority (GLA) following the abolition of the Greater London Council in 1986. The 2004 London Plan had a section addressing gender in a chapter on London's diversity (Mayor of London, 2004: 72). These broad outlines were further elaborated in 2007 in a Supplementary Planning Guidance document (SPG) (Mayor of London, 2007). The definition of "protected characteristics" was extended to travellers, refugees and asylum seekers. The SPG is aspirational in tone, but does not consider intersectionality between the protected groups and lacks detail on specific policies to be adopted by the individual local authorities. Nevertheless, it remained as guidance to subsequent redrafting and iterations of the London Plan.

In addition to the London Plan, the Mayor has the ultimate say over any major development of 15,000 m^2 or more outside of central London and any housing development with over 150 residential units. While the subsequent changes and alterations to the London Plan between 2011 and 2017 could be characterised as primarily gender blind, this should be viewed in the context of wider changes to national and local policies in the period dating back to the late 1990s. In this period some of the principles of spatial development policies that are perceived to be in accordance with gender-sensitive criteria came to be adopted as mainstream.

A senior female planning officer in one of London's development corporations commented in an interview with the UK author:

> I'm not entirely sure if we've got a specific section in here [planning document produced by the corporation] on gender or not, I think it's just about making places accessible to everyone and good places to live in, with local employment, with good transport and these are all the things that will make everybody's life great, easy, not just women.

The Mayor set out a London-specific set of residential design standards and this includes criteria for the *quantum* of public and private spaces that should be within either the curtilage or easy reach of each dwelling (Mayor of London, 2012, 2016).

Following the election of a Labour Mayor in 2016, a new draft London Plan was produced. This brought inequalities to the fore in the concept of "good growth". The new draft London Plan (Mayor of London, 2017) has been subjected to an "Integrated Impact Assessment" (IIA) (Arup, 2017). This is a scrutiny that combines four assessments: strategic environment, equalities, health and well-being, and crime and safety. The IIA was carried out by an external consultancy, Arup, and the results fed into an iterative dialogue with the planners together with other consultations at the GLA. The impacts are assessed with regard to their short-, medium- and long-term effects, whether they are direct or indirect, temporary or permanent and whether they are felt at the local, Greater London, wider region and/or global scale. The strategic options for "good growth" are assessed in a similar fashion. This exercise yielded an enormous amount of information. It should be noted that, at regional level, it is the projected impacts that are being assessed in a remote "expert" process.

The indicators that will be used for monitoring and which relate directly to gender are somewhat limited:

- Reduction in the difference in life expectancy between those living in the most and least deprived areas of London (shown separately for men and women)
- Increase in the rate of employment of lone parents (who are mainly female) relative to the average for England and Wales and increase in the employment rate of black and minority ethnic Londoners

Another indicator in which gender criteria are implicit lies in increasing the relative modal share of public transport.

Overview of gender sensitivity in urban development in Vienna – STEP 2025 Urban Development Plan Vienna

Vienna has more than 25 years of experience in the implementation of gender mainstreaming in spatial planning and development. Initially, the topic was framed as equitable city development for women. The starting point was an exhibition in 1991: "Who owns public space – women's everyday life in the city". The Women's Office (Municipal department 57), initiated in 1991, had planning issues as its central focus. In 1998 a "Coordination Office for Planning and Construction Geared to the Requirements of Daily Life and the Specific Needs of Women" was established in the Executive Office for Construction and Technology of the City of Vienna.

From 2000 onwards gender mainstreaming became a cross-cutting strategy for the whole municipality as well as in spatial planning and development. Between 2005 and 2010 all the departments connected with spatial development in the City of Vienna had to conduct gender mainstreaming pilot projects as a first phase. Some 60 pilot projects at different planning levels were implemented, ranging from small urban spaces to a pilot district, from public space, parks and playgrounds, urban design, housing, public purpose buildings and lighting. In the second phase, the focus was on structural and more systematic implementation by knowledge transfer, evaluation and reporting procedures. The approach was process orientated, integrated in an inter- and transdisciplinary discourse facilitated by the personal continuity of the leading gender planning expert (see Irschik and Kail, 2013).

Various planning manuals and guidelines were developed to support municipal planners and external planning consultancies to incorporate gender mainstreaming into their work. These range from criteria lists for urban development plans and checklists for streetscape planning to planning recommendations for gender-sensitive park design or lists of criteria to evaluate the gender sensitiveness and everyday needs of orientation projects inquiring into housing subsidies. All of these were derived from the implementation of these projects. These activities at different levels of planning provided a valuable resource for the integration of gender perspectives into urban development documents.

Vienna has had strategic guidance for urban development since 1980. Approximately every ten years the City of Vienna makes a new urban development plan. In 1994 women-equitable city planning was part of the urban development concept (Stadtentwicklung Wien, MA 18, 1994), for the first time. In 2005 gender mainstreaming was introduced as one of the basic principles and as a strategy to take the needs of different user groups into account. A discrete chapter showed the benefits of the gender perspective in urban planning. Gender was also mentioned in every relevant section (Stadtentwicklung Wien, MA 18, 2005).

In 2014, STEP 2025 was published. Gender mainstreaming is explicitly mentioned several times within different planning tasks – starting with the objectives and context for urban development in Vienna: "An active municipal policy on behalf of gender equity supports the creation of equitable life conditions for women and men. Gender-sensitive planning methods thus take conscious account of different interests and needs" (Vienna City Administration MA 18, 2014: 9). Gender mainstreaming is integrated as a cross-sectional issue. Social inclusion is a main principle for the concept of integrating aspects of gender and diversity. Thematic concepts on "Mobility", "Green and Open Spaces" or "Public Space" followed and specified the contents of the urban development plan including gender aspects and instruments such as the fairness check to assess the proposed measures of the concepts at the end of the process.

As a contribution to STEP 2025 a manual on gender mainstreaming was published including guiding principles and criteria for gender-sensitive urban development (Damyanovic, Reinwald and Weikmann, 2013). The Manual "Gender Mainstreaming in Urban Planning and Urban Development" was developed parallel to the STEP 25 process. It explains the mainstreaming strategy as a basis for gender-sensitive planning procedures which foster the integration of gender equality at all stages of the planning process, ranging from the formulation of objectives and the planning of measures through to their implementation and evaluation. In addition, it is important to question planning models and the values and attitudes underlying the planning philosophy and planning culture with regard to gender equality. Beside the objectives, profiles of people's changing needs throughout their lifecourses, in particular, detailed planning indicators for quality and methodological approaches developed over 25 years are presented. Densification is an important objective in STEP 2025. The term "high-quality density" was used in the Gender Manual to stress the fact that high density needs adequately supplementing with open spaces and green areas and careful urban design that considers shading effects for open space and noise reduction.

Differences and similarities

In London gender is regarded as a "protected characteristic" and is therefore not explicitly highlighted as a concern for planning apart from in a limited manner. Vienna, by contrast, has an explicit commitment to gender sensitivity and the appointment of gender experts. The extent and thoroughness of the City of Vienna's commitment to gender mainstreaming is impressive. The Viennese experience and history of implementing gender mainstreaming far outpaces that of London and there is much to be learned from its 60 pilot projects, spatial development plans, manuals and guides. In Vienna this learning was necessary to "translate" the strategic approach of gender mainstreaming into urban planning and development.

As Greed and Reeves (2005) and Berglund (2013) have commented, gender gets "lost" in the UK planning system, but some of the objectives that are identified as gender sensitive in Vienna are given priority in the London Plan and are implemented at borough level under the concept of environmental sustainability. Implicit gender-sensitive objectives also overlap with the concept of "health and well-being" in the UK. The nature of these gender-sensitive objectives and concepts are discussed in the next section.

Assessment criteria for gender sensitivity in urban development for the case studies

There are many checklists and lists of criteria to assess gender-sensitive urban development from different points of view, e.g. transport planning, housing, park design and different theoretical backgrounds (e.g. Knoll 2008; Damyanovic, Reinwald and Weikmann, 2013; Sánchez de Madariaga, 2013; Damyanovic, 2016). However, there is no framework commonly agreed upon for strategic objectives of gender-sensitive urban planning and development, although many research projects on individual elements have been conducted (e.g. Gutmann and Neff, 2006; Zibell, 2006; Senatsverwaltung für Stadtentwicklung, 2011; Col·lectiu Punt 6, 2015).

The Everyday Life approach (Lefebvre, 1971) is a central concept within gendered planning for the assessment of the built environment and urban development (Damyanovic and Zibell, 2013). Everyday spatial practices are scrutinised for the possibilities offered to diverse groups to claim city spaces and to make differences visible (Beebeejaun, 2017). The Everyday Life approach "is more than a theory or a critique. It is also a vision of a more harmonious society in which people are at the centre of all concerns rather than the pursuit of the quick economic fix" (Horelli, Booth and Gilroy, 2000: 9).

The dependency of people on their local environments varies according to life circumstances. Taking into account these different needs and requirements is another commonly agreed objective of gender-sensitive urban development and is underpinned by gender-differentiated statistics and data (Damyanovic, 2007). Closely linked to this is the objective of a fair distribution of space and resources. The unequal distribution of paid and unpaid work, services and amenities offered by cities leads to unequal living conditions. Gender-sensitive urban planning and development has to address this unequal distribution and simultaneously design urban structures which facilitate the everyday routines of people to manage caring and domestic responsibilities. An attractive and safe living environment which allows women and men to move freely through the city at any time is central to a good quality of life. This is especially so for people who are more dependent on their local environment (e.g. elderly women and men, those with caring obligations, children). These objectives are all supported by the "Right to the City" approach, which also implies the equal representation and participation of women and men in decision-making processes in urban planning and development (Mitchell, 2003; Fenster, 2005; Fainstein, 2010).

Some (more or less) "mainstream" planning models support these objectives and approaches to gender-sensitive urban development. A polycentric structure with a mix of uses and environmentally friendly public transport infrastructure supports a city of short travel distances and thus the quotidian lives of women and men. The corollary of such a structure is a decentralised, even distribution and provision of essential commercial and social infrastructure. High-quality public and green spaces as well as a safe and barrier-free city are key features of an attractive living environment and provide the basis for women and men to take part in city life (Mitchell, 1996; Wiener Stadtgärten, MA 42 and Stadtbaudirektion Wien, 2005). The provision of a wide range of housing types with flexible layouts is another leading principle of mainstream planning which supports more diverse urban areas and allows for changing needs.

Based on these objectives and approaches of gender-sensitive urban development and "mainstream" planning models the following criteria for assessing the case studies were chosen:

- Polycentric structure – development of local centres, decentralised concentration of service and infrastructure facilities close to high-level public transport stops
- Mix of uses – mix of residential buildings, workplaces, shopping and leisure facilities, ease of access to employment for all genders with regard both to range of employment opportunities and transport links
- Provision of a wide range of housing types – manifest range of residential typologies and layouts complying with various tenancy or ownership structures, financing and cost requirements

- High-quality public and green spaces – adequate private and communal open spaces for everyday tasks, children's play and leisure, differentiated functions of various open space types (e.g. open/green spaces near the home, streetscapes or parks and squares) ranging from totally private to totally public
- Provision of essential everyday life and social service infrastructure – social infrastructure facilities for all population groups, high-quality planning for kindergartens and schools with respect to their location, availability of easily reachable basic shopping outlets and service providers in the neighbourhood
- A city of short travel distances and (environmentally friendly) public transport infrastructure – a tightly knit route network especially for pedestrians and cyclists, development oriented towards high-level public transport stops like metro or tramway, quality of streetscape
- Safe and barrier-free city – effective and clear-cut spatial orientation and social control in the neighbourhood, use of street level/ground floor, support of good neighbourly relations by e.g. neighbourhood management
- Equal representation and participation of women and men in planning and decision-making processes

Case studies on gender sensitivity in urban development

The implementation of the criteria for gender sensitivity in urban development master plans are described and compared for two case studies, London (Barking Riverside) and Vienna (Aspern Seestadt).

Case study London – Barking Riverside

Barking Riverside is a major development site of some 180 ha in the east of London. It is located in a low income area which was identified as an area for expansion known as the Thames Gateway. The development pressures are less acute than in other parts of London and local property prices are lower (LB Barking and Dagenham, 2010). The information that follows is taken from the Design and Access Statement for the planning application for the master plan (Greater London Authority, 2016).

General description

The site was the former home of three power stations and a landfill site. It is bordered on one side by the River Thames and on the other by a major strategic link road. The major area of the site is accessed by one road and it is only the proposed extension of the public transport system that has released it for regeneration (Greater London Authority, 2016). A master plan for one land parcel was prepared in the late 1990s and approximately 800 homes were completed by a joint venture public-private partnership. A master plan for the whole site was

Figure 4.1 Proposed Masterplan for Barking Riverside 2016.
Source: © Lifschutz Davidson Sandilands.

approved in 2009 for a mixed-use sustainable community which would accommodate 10,800 homes. By 2016 only the first phase of that master plan had been completed, with 686 dwellings and a primary school. This first phase was designed to high eco-standards and won two awards. However, the residents were left without "a café, a pub or a doctor's surgery" and had to wait years for a convenience store to open (Burrows, 2015).

Later phases were delayed through lack of investment for the proposed extension to the public transport system. In 2016 a new station on the London Overground, Barking Riverside, was approved and is to be completed by 2020. A new partnership has been formed between a housing association, London and Quadrant New Homes, and the GLA. A revised master plan, which substantially rethought the townscape, site layout and architectural forms, was prepared by Lifschutz Davison Sandilands and approved in 2016. Following this, work started in earnest on design and construction. In the course of this extended time period the context for development changed. London's population grew by 25 per cent, 1.7 million people, between 1997 and 2016 but its housing stock only increased by 15 per cent, approximately half a million homes (Mayor of London, 2018: 28–29). Furthermore, intense property speculation fuelled a housing crisis. In consequence, the Mayor's housing strategy and draft London Plan gives high priority to an absolute increase in the number of new residential units.

Gender sensitivity in the development concept of Barking Riverside

The marketing of the proposal undermines the concept of Greater London as a polycentric city through its insistence that Barking Riverside will be approximately 25 minutes away from the centre of London by a combination of the Overground and Underground railway systems. This implies that its accessibility is important for residents who wish to work in central London's financial district, the City. More support for a dispersed urban structure is provided by the Overground extension, which places the new development only eight minutes from its nearest town centre, Barking.

Barking Riverside is residential led, although the developers and architects aspire to creating "the setting for a balanced and inclusive community" (Greater London Authority, 2016). Residential uses will occupy 84 per cent of the floor space of the site when all four phases are completed, 7 per cent will be allocated to schools, 3.5 per cent to car parking, 2 per cent to retail, cafes and bars, and 1 per cent to offices. The remaining 1 per cent is allocated to leisure uses, a new GP surgery and health centre and workshops. Two possible sites have been identified for a major new food retailing store. There are no proposals for new employment sites, but there is a new environmental technology park adjacent to the development. A district centre is proposed around the transport hub, together with a public square. A minor centre plus some small-scale retail is proposed at the other end of the main spine route through the development.

Figure 4.2 Proposed Masterplan (3D view) for Barking Riverside 2016.

Source: © Lifschutz Davidson Sandilands.

The proposals offer a careful mix of dwelling sizes, with a greater emphasis on three- and four-bedroom sized units in the early phases. High towers are only included in the final fourth phase. The scheme, although medium density, is of a human scale with carefully proportioned perimeter blocks. The mix of tenures is more controversial by London standards with a baseline of only 5 per cent of units being offered at a "social rent" that is the same as council housing, approximately 25 per cent of market rents. The bulk of the 50 per cent "affordable" units are shared ownership units.

The provision of publicly accessible green space is impressive. Two major parks are proposed. The central spine road has a green central strip 12 m wide as its boulevard. The master plan sets out guidance for high-quality public spaces with careful attention paid to a hierarchy of streets and paths. A considerable degree of thought has been put into the guidance for the different types of streets and spaces, ensuring that different consultants can maintain a coherent and consistent quality. The district centre incorporates a main square and riverside transport landing point, making the most of the potential links the Thames can offer between its two sides. Barking Riverside was one of the pilot projects for a European-funded research/action project, TURAS (2016). The European funding assisted in developing new ideas to establish biodiversity, to develop resilience with regard to flooding, and to experiment with green roofs. The findings were fed into the revised master plan process.

The proposal for a mix of uses and social infrastructure covers basic needs for shopping and the possibility of some cultural and social uses. The architectural forms proposed in the master plan allow some flexibility at ground-floor level to permit alternative uses to residential. In the UK many services are outsourced or privatised, or are run by agencies with only an "arm's length" relationship to local municipalities. Therefore negotiations take place over the provision of public services such as police stations, health centres and schools. Austerity programmes after the 2008 banking crisis have led to 40 per cent cuts in local council budgets and the closure of many facilities such as libraries and children's services.

Two new dedicated, high frequency bus routes, branded as the ELT (East London Transit) will provide links to Barking town centre on one side of the development and Dagenham town centre on the other. A new road connection will also be provided to the eastern side of the site. The central spine route running through the development will be car free. The ELT will run along this route, at six-minute frequencies. The Mayor's housing design standards require that bike storage be provided for each block, and bike storage will also be provided adjacent to the transport hub. Non-motorised routes are provided throughout the development, but each apartment block will be accessible by taxi. The majority of the development will be within a ten-minute walking distance of the district centre and schools.

New stations and public buildings are built to the requirements of the Disability Discrimination Act of 1995. Residential design standards require 10 per cent of new homes to be Lifetime Homes that can easily be used or converted to use by people in wheelchairs. On the other hand, the site itself is set on a slope and the architects have exploited this in a sensitive manner to create areas of different character. Interestingly, the Design and Access statement for the master plan does not elaborate on the accessibility of the hillier parts of the site for residents with restricted mobility.

Barking and Dagenham Council consulted a planning forum in the preparation of the first master plan. Prior to the initial phase of development, a Community Interest Company (CIC) was set up in 2009 to represent the community and to hold and manage its assets, i.e. the green spaces and community buildings. A new partnership has been formed with the National Health Service to pioneer Barking Riverside as a Healthy New Town. Initially the CIC will be run by the development company and the local authority until there is a sufficient number of and capacity amongst the local residents to hand the CIC over to resident control (Barking Riverside London, 2018).

Case study Vienna – Aspern Seestadt

Aspern Seestadt (Aspern urban lakeside) with 240 ha is one of the major development areas in Europe, situated in the north-east of Vienna.

General description

The development site is a former airfield that opened in 1912 and closed in 1971. In the 1980s, an automotive engine plant was built on the southern part of the area and the airfield was used as a training area for learner drivers.

Vienna experienced strong growth from the 1990s onwards following the opening of the "Iron Curtain". The population rose from 1.5 million in 1991 to 1.9 million in 2017. This put pressure on urban development and the housing market. The trend is expected to continue, to 2 million people in the late 2020s. Therefore Viennese urban development focuses on developing new urban quarters, especially on former brownfield or railway sites. Aspern Seestadt is the largest and is supposed to provide approximately 10,500 flats accommodating more than 20,000 people, and 20,000 jobs.

The planning process started in 2003 with a project team from the City of Vienna and a strategic environmental assessment. In 2004 the "Wien 3420 Aspern Development AG" was founded, a private-public partnership.

A two-phase, European-wide competition for the urban design took place. The winners in 2005 were the Swedish office Tovatt Architects & Planners together with the German developer office N+ Objektmanagement GmbH. They were commissioned with the development of the master plan (Tovatt Architects & Planners and Projektteam Flugfeld Aspern, 2007).

In 2009 guidelines for the public spaces and streetscapes, produced by Gehl Architects ApS, were published (Gehl Architects ApS, 2009). This had been also the subject of a competition. In 2010 the artificial Lake Aspern (5 ha) was excavated and in 2011 the first streets and basic technical infrastructure were

Figure 4.3 Masterplan for Aspern Seestadt.

Source: © Wien 3420 aspern development AG.

completed. In 2013, one year before the first inhabitants moved to the new urban district, the extension of the metro system opened.

2,900 flats were built accommodating 6,000 residents and 1,500 workspaces have been created as of May 2018 (Wien 3420 aspern development AG, 2018). The master plan was updated twice, in 2012 and 2018. The basic concepts such as the ring road and the open-perimeter block development have been retained throughout. Now a second section close to the underground station is under construction, also including high-rise buildings of mixed use.

Gender sensitivity in the development concept of Aspern Seestadt

The main principles for the development were to create urban density and therefore achieve a city of short travel distances with a strong focus on the quality of public space. The dimension of a small town was intended to help to achieve urban qualities in a peripheral location, surrounded by a neighbourhood characterised by agricultural land use, single-family houses and a limited number of multi-storey buildings of moderate density.

Therefore Aspern Seestadt is planned as a local centre and mixed-use area – living, working, production, research, culture and social infrastructure, following the idea of a polycentric city structure. Although the goal was to reach an equal share of residents and employment opportunities, for the first phase of development this proved difficult. Due to its peripheral location and the lack of a motorway connection, and despite the fact that the City's agency for economic support built a technology centre, fewer firms and enterprises than expected moved to Aspern Seestadt in the initial years. The City strengthened the mixed-use nature of development by relocating municipal departments to increase office floor space; the technology centre is expanding.

The aim is to achieve a mix of social (developed by cooperative developers/associations) and privately financed housing. More than half of the houses built up to 2018 were subsidised by the City of Vienna. Five co-housing projects are concentrated in one central block and two students' hostels have been built, one for temporary use.

Figure 4.4 Aerial view of the southern part of Aspern Seestadt.

Source: © City of Vienna, MA 18/C. Fürthner.

The lake and parts of the lakeside park were built before the residents moved in. Two additional parks were developed and possibilities for urban gardening were offered. A strong emphasis was put on the design of public space and the streetscapes. The residential roads form part of the open space network and many of them are car free.

The urban design includes a small local centre, which was already constructed for the first phase, with shopping facilities covering everyday needs. Aspern Seestadt has the first managed shopping street in Austria. The developers had to provide business premises and shops on the ground floor. An agency was founded, part-owned by 3420 AG and part-owned by one of the major companies that runs grocery stores and shopping malls in Austria. The agency is responsible for the management of these ground-floor zones, to ensure a mix of shops. In the first phase, 3,000 m^2 of business premises for 13 shops, food outlets and small businesses have been completed. Additionally there are a neighbourhood management, medical centre, pharmacy, kindergartens, a primary school, grammar school and small private childcare facilities (children groups).

The basic urban outline is built of open-perimeter blocks which are quite small (average length 84 m, average width 57 m) compared to other urban brownfield development projects in Vienna (average length 126 m, average width 79 m) (Stadtentwicklung, MA 18, 2017: 192). This provides a pedestrian route network which is very dense, and monotonous roads are avoided.

The final modal split is designed to reach 40 per cent of cyclists and pedestrians, 40 per cent public transport and 20 per cent motorised private transport (by comparison, Vienna achieves 33 per cent cyclists and pedestrians, 39 per cent public transport and 27 per cent motorised private transport). To reach this aim car parking is only included in some of the perimeter blocks. The idea behind this is that the trip from home to the car is as far as the trip to the next public transport station (max. 300 m for both). This helps to reduce car traffic, augment pedestrian routes and support street life. Also bicycles, e-bikes and cargo bikes are available to rent. Major investment into the extension of the metro system enables a journey to the city centre within 20 minutes. A bus route, leading through the first quarters to be completed, connects the stations and improves connectivity.

A neighbourhood management team accompanies new residents on arrival, informs them about current developments and supports the active co-creation of living together, for instance through the use of the "white spots" left undesigned in public space for future use.

The use of the ground floor and the perimeter block structure also supports safety in the streets by providing "eyes on the street" (Jacobs, 1961). However, until all sections of the development are complete, some residents need to walk for quite a long distance without buildings to reach the underground station, although pedestrian routes are well lit.

A monitoring study showed that all these measures have supported satisfaction with the living environment, with 82 per cent of the inhabitants responding reporting that they enjoy living in Aspern Seestadt (Reinprecht *et al.*, 2015).

At the start of the development process there was a public inquiry for residents in the nearby neighbourhood, most of who lived in single-family houses. Three representatives were elected to accompany the master plan process, including one woman.

The incorporation of people living in the neighbourhood, their information and participation from the very beginning had no specific gender aspect, but offered deep engagement. A municipal department and later the development agency "branded" the unfamiliar site of the former airfield in an innovative, socially sensitive way. For example, unemployed young people were involved in constructing outdoor furniture that could be used in future sites or the establishment of a youth centre. Many cultural activities have also taken place.

The jury for the tender for the master plan was composed of 15 people, four of whom were women including one female resident representative and the head of the political district board. One of the two CEOs of the development agency was a woman until 2015. She was responsible for urban design and planning and had a decisive role in shaping the development process. On her initiative almost all of the streets, parks and squares are named after women. This is an impressive signal on the symbolic level, given that only 3 per cent of streets in Vienna are named after women (42 per cent after men) (Rathkolb *et al.*, 2013). The interdisciplinary advisory board created in 2010 to accompany the development process has equal representation of women and men.

Gender mainstreaming in Aspern Seestadt development process

The development of a master plan is a long process. Explicit gender mainstreaming activities were included in many development phases, but this was not a systematic roll-out including all planning activities at all levels. Nevertheless, a full range of different gender planning actions was deployed and became visible at Aspern Seestadt.

At the beginning the tender for the development of the master plan had no specified gender aspects. The gender planning experts from the Coordination Office were involved in the process of master planning as the Municipal Department 21, "District Planning and Land Use", had to propose a gender mainstreaming pilot project out of their annual work programme. This decision provided the entry point for explicit gender planning activities. A study was commissioned by the municipal department, which found the most gender-relevant aspects to be accessible on foot, adequate social infrastructure and the functional qualities of semi-public and public open spaces to accommodate different users (Gutmann and Neff, 2006). The study identified social infrastructure elements, made proposals for their location and developed a trip chain assessment, testing eight gender-sensitive daily life patterns for four housing sites, using the proposed locations of social infrastructure and the proposal for the land-use plan to test the "city of short travel distances".

A gender planning expert was involved by the development agency in the preparation of the neighbourhood management scheme and participated in the process for the modification of the northern part of the master plan. Gender planning experts were involved as jury members in the developer competition for 760 flats, and in one of the two park competitions. For the competitions for the educational buildings a gender-sensitive prequalification took place. The guidelines of the City of Vienna for gender-sensitive park and playground design were part of the tender requirements in both park competitions. In all the competitions the experience of former gender-planning activities, such as the gender pilot project on school campus planning, gender mainstreaming within developers' competitions and gender-sensitive park design were a prerequisite for formulating adequate gender positions.

Differences and similarities

The comparison of the master plans for Aspern Seestadt and Barking Riverside reveals many shared characteristics. Both rely on a perimeter block approach to urban design, both place emphasis on the design and accessibility of public spaces, both have a coherent and legible street system and both certainly aspire towards a sustainable and environmentally friendly transport system. Both developments have ample green space and aim for a high quality of biodiversity, combined with leisure and recreational use (see Table 4.1).

The major difference between the two developments lies in the quantity and quality of social and commercial infrastructure. Aspern Seestadt will have major employment functions, a strong shopping street and a multitude of social and commercial premises. By contrast, the mix of uses proposed for Barking Riverside appears to be rather meagre. As such, the goal of creating a rich and diverse neighbourhood seems unlikely to succeed. This lack of infrastructure arises as a consequence of a number of factors, including a severe housing shortage and house price speculation, the UK's insistence that infrastructure costs should be paid for by substantial contributions from developers, combined with cuts in social provision.

There is also a difference between the planning processes. The development of Aspern Seestadt included several explicit gender planning activities, even if this was not the case from the very beginning. Beside the specific gender pilot project and the involvement of gender experts in some of Aspern Seestadt's specific planning activities, gender expertise which was already firmly established in general mainstream planning procedures was well integrated.

Discussion and conclusion – gender sensitivity in urban development and its "transmission" into mainstream planning

The comparison between the two case studies has highlighted the extent to which gender inequalities are obscured as a specific issue to be addressed in the planning system for England and Wales. Here a distinction should be made between gender equality and diversity (Malik, 2018). The inclusion of all the "protected characteristics" in the London Plan's Integrated Impact Assessment

Table 4.1 Comparison of the masterplans Barking Riverside and Aspern Seestadt

Assessment criteria	*Barking Riverside*	*Aspern Seestadt*
Polycentric structure	Not fully realised as lack of local employment on site	Yes, serves as a new centre also for surroundings, several local sub-centres
Mix of uses	This is the aspiration but housing shortages may crowd out other ground-floor uses dispersed through the neighbourhood	Mix of housing and employment, but hard to achieve initially (to date the number of workplaces has not reached the target)
Wide range of housing types	Yes, many proposed, low provision of social rental housing	Yes, publicly and privately financed, high levels of social rental housing
High-quality public and green spaces	Yes, two principal parks and other spaces	Yes, different forms of parks, places and semi-public open spaces
Safe and barrier-free city	Yes, although central boulevard may pose challenges after dark	Yes, although in the first phase the route to the metro station was without "eyes on the street" due to a lack of buildings, now in the second phase these buildings are under construction
Everyday life and social service infrastructure	Mainly to be provided by commercial and third sector, low level of provision compared to established neighbourhoods	Yes, broad range of public social infrastructure as well as shops, 4 m high ground floors in the main streets, higher level of provision than in the surroundings
Short travel distances and public transport	Good transport links to Barking and central London.	Yes, dense network of pedestrian routes and public transport system including extension of metro line
Equal representation and participation	Aspiration for local, sensitive governance with CIC	Not from the very beginning but in some stages

results in a complex assessment document of bewildering length (Arup, 2017). While an attempt to be inclusive is laudable, it has led to the driving concept behind a gender-sensitive approach, to reconcile the needs of everyday life, becoming eclipsed in practice.

In Austria the integration of women- and gender-issues in strategic high-level policy documents at federal and state level is made in accordance with the top-down approach to an overall mainstreaming strategy. The situation in spatial development is different. Due to the decentralised federal planning system and the multiple planning levels and responsibilities, implementation varies greatly between the federal states, cities and municipalities.

Also in Vienna, as the most advanced in regard to gender planning issues, the implementation of a gender mainstreaming strategy was not the result of a consistent top-down process, trickling down the different planning levels, but was developed in parallel at all levels and with a broad thematic focus, with the Coordination Office and a leading councillor playing important roles. A range of pilot projects and pilot processes was implemented, with the result that many of the methods or contents developed have been picked up by the mainstream.

The approach of the City of Vienna demonstrated that transmission into mainstream planning is possible, but has certain prerequisites and may take a number of years. Every planning level and planning task had to learn what the implementation of gender mainstreaming meant by starting with pilot projects. This dissemination had to be tailored to the specific urban and social conditions, the administrative structure of the city and – most importantly – to the different planning levels and areas. But there are still gaps in techniques and in the operationalisation of gender mainstreaming at the different levels of spatial planning and development in Vienna. Favourable conditions for transmission are context dependent. Role models from one city, one planning level or sphere of planning cannot just be "copied and pasted".

Gender reaches the mainstream at different planning levels and fields in different ways

This transmission or dissemination through different scales of planning and planning activities is a key success factor in Vienna. It has been possible to introduce many gender mainstreaming criteria, with considerable impact especially in urban design, housing, park design and the transport and mobility sector. In public parks, which proved to be the role model of successful mainstreaming, the public sector is normally the landowner and therefore has the power of interpretation and definition to provide an adequate offer for different user groups. A similar observation can be applied to street design and public space in general. Influence decreases in the area of local services because this is largely left to the market economy.

By contrast in London, it is the sustainability agenda that is encouraging a city of short travel distances and a high level of public transport. Design review panels at borough level help to assess master plans and development schemes, thereby encouraging a high quality of street design that addresses the needs of pedestrians and disabled people. The extent to which park design will accommodate the needs of all users is dependent on the skills of the officers involved in writing the brief and the landscape architects commissioned. In the absence of a set of guidelines explicitly addressing gender criteria and a procedure for implementation, as in Vienna, gender sensitivity cannot be guaranteed.

Loss of influence due to PPPs and reduced public budgets affects the quality assurance associated with gender mainstreaming

Cities are losing more and more of their influence over urban planning and development, which is crucial in terms of the quality assessment required by gender mainstreaming. The City of Vienna has (or had) a strong influence on the quality of its subsidised housing throughout and far less impact on privately financed residential construction. But the city is losing influence over the implementation of gender mainstreaming in the housing sector, as the proportion of private and international investment constantly increases.

This process has been noted in London, where social provision has to be negotiated with different bodies and agencies, and commercial services – such as shops and cafes – depend on the willingness of investors to provide them (Manzi *et al.*, 2010). It is frequently stipulated that essential infrastructure, such as the new station in Barking Riverside, be provided from the developer's profit, from "planning gain". In Vienna, as was already the case in London, there is an increasing emphasis on using planning gain to require investors to finance public facilities such as kindergartens, schools or parks, using the new instrument of the building code, the so-called "urban development contract", for this purpose.

Has gender-sensitive planning arrived in the mainstream?

High-quality mainstream planning incorporates many gender-sensitive qualities. In addition, methodological approaches such as social space analysis have been (partially) adopted in the mainstream in Vienna. This has been aided by a wider recognition of the importance of designing streets for people, a revaluation of the properties of historical cities and a promotion of public space (see for example Gehl, 2010; Madanipour, Knierbein and Degros, 2014). The concepts of sustainability have ensured that many of the criteria developed for gender-sensitive design extend into spatial planning objectives.

The evaluation of planning processes and outcomes is difficult: the Viennese example shows that the more effectively gender flows into the planning mainstream, the more "invisible" it becomes. The outcome is that it is difficult to tell what resulted from explicit gender planning activity, and what from implicit gender-sensitive planning or just high-quality mainstream planning. This becomes more visible only if you analyse the process carefully and in detail.

Most of the time, planning deals with conflicts of goals. In the "planning mainstream", these goals are just discussed as functional ones. Gender planning makes it possible to support socially sensitive decision-making in a systematic way. This is also valuable for the prioritisation of measurements. Gender-sensitive planning methods used in addition to high-quality mainstream planning, like the Viennese fairness check, add another level and considerably increase the complexity involved in relating different goals and functional aspects of varying degrees of importance to different needs and social groups. Nevertheless, this chapter has tried to demonstrate the value of such an effort.

Note

1 The local development concepts of the following cities were analysed: Vienna, Graz, Linz, Salzburg, Innsbruck, Klagenfurt, Villach, Wels, Bregenz, Eisenstadt as of 2016.

References

Arup (2017). *Greater London Authority. London Plan. Integrated Impact Assessment. Consultation Document.* [Online] London: Ove Arup & Partners. Available at: www.london.gov.uk/sites/default/files/draft_london_plan_iia.pdf [Accessed 7 Apr. 2018].

Barking Riverside London (2018). *Community.* [Online]. Available at: https://barkingriverside.co.uk/community/ [Accessed 7 Apr. 2018].

Beebeejaun, Y. (2017). Gender, urban space, and the right to everyday life. *Journal of Urban Affairs*, 39(3), 323–334.

Berglund, E. with Wallace, B. (2013). Women's design service as counter-expertise. In: I. Sánchez de Madariaga and M. Roberts (Eds.), *Fair shared cities. The impact of gender planning in Europe.* Farnham: Ashgate, 249–264.

B-VG, Bundes-Verfassungsgesetz BGBl. Nr. 1/1930 idF BGBl. I Nr. 194/1999 (Federal Constitution Law). [Online]. Available at: www.ris.bka.gv.at/GeltendeFassung.wxe?Abfrage=Bundesnormen&Gesetzesnummer=10000138 [Accessed 9 May 2018].

Burrows, T. (2015). No café, no pub, no doctor in London's most isolated suburb. *Guardian*, [Online] 17 Aug. 2015. Available at: www.theguardian.com/cities/2015/aug/17/no-cafe-pub-doctor-londons-most-isolated-suburb-barking-riverside [Accessed 7 Apr. 2018].

CLG – Department for Communities and Local Government (2012). *National Planning Policy Framework.* [Online] London: Department for Communities and Local Government.

Col·lectiu Punt 6 (2015). *Women working. Urban assessment guide from a gender perspective.* [Online]. Available at: https://issuu.com/punt6/docs/ww_issuu_simple/114 [Accessed 9 May 2018].

Damyanovic, D. (2016). *A gender-sensitive approach in landscape planning. Theoretical and methodological prerequisites and findings of basic and applied research to develop further a gender-sensitive approach in landscape planning.* Professorial Thesis – University of Natural Resources and Life Science Vienna, Vienna.

Damyanovic, D. (2007). *Landschaftsplanung als Qualitätssicherung zur Umsetzung der Strategie des Gender Mainstreaming (Landscape planning as quality assurance for the implementation of the strategy of gender mainstreaming).* Vienna: Guthmann-Peterson.

Damyanovic, D., Reinwald, F. and Weikmann, A. (2013). *Gender mainstreaming in urban planning and urban development.* Werkstattbericht, Nr. 130A. [Online] Vienna: Urban Development Vienna, Municipal Department 18 (MA 18) – Urban Development and Planning. Available at: www.wien.gv.at/stadtentwicklung/studien/pdf/b008358.pdf [Accessed 9 May 2018].

Damyanovic, D. and Zibell, B. (2013). Is there still gender on the agenda for spatial planning theories? Attempt to an integrative approach to generate gender-sensitive planning theories. *disP – The Planning Review*, 49(4), 25–36.

Fainstein, S. (2010). *The just city*. Ithaca, NY: Cornell University Press.

Fenster, T. (2005). The right to the gendered city. Different formations of belonging in everyday life. *Journal of Gender Studies*, 14(3), 217–231.

Gehl, J. (2010). *Cities for people*. Washington, Covelo, London: Island Press.

Gehl Architects ApS (2009). *Partitur des öffentlichen Raums. Planungshandbuch Aspern Seestadt (Score of public space. Planning manual Aspern Seestadt)*. Werkstattbericht, Nr. 103. [Online] Vienna: Urban Development Vienna. Available at: www.wien.gv.at/stadtentwicklung/studien/b008068.html [Accessed 9 May 2018].

Greater London Authority (2016). *Planning report D&P/0150c/01–22 March 2016. Barking Riverside – planning application no. 16/0013/OUT*. [Online]. Available at: www.london.gov.uk/sites/default/files/PAWS/media_id_314010/barking_riverside_report.pdf [Accessed 7 Apr. 2018].

Greed, C. and Reeves, D. (2005). Mainstreaming equality into strategic spatial policy making. Are town planners losing sight of gender? *Construction Management and Economics*, 23(10), 1059–1070.

Gutmann, R. and Neff, S. (2006). *Gender Mainstreaming im Stadtentwicklungsgebiet Flugfeld Aspern. Begleitende Expertise zum Masterplan (Gender mainstreaming in the urban development area Aspern airfield. Accompanying expertise to the master plan)*. [Online] Vienna: Wohnbauforschung. Available at: www.wien.gv.at/stadtentwicklung/projekte/aspern-seestadt/pdf/studie-gendermainstreaming.pdf [Accessed 9 May 2018].

Horelli, L., Booth, C. and Gilroy, R. (2000). *The EuroFEM toolkit for mobilizing women into local and regional development*. Helsinki: Helsinki University for Technology.

Irschik, E. and Kail, E. (2013). Vienna. Progress towards a fair shared city. In: I. Sánchez de Madariaga and M. Roberts (Eds.), *Fair shared cities. The impact of gender planning in Europe*. Farnham: Ashgate, 193–230.

Jacobs, J. (1961). *The death and life of great American cities*. New York: Random House.

Knoll, B. (2008). *Gender Planning. Grundlagen für Verkehrs- und Mobilitätserhebungen (Gender planning. Principles of traffic and mobility surveys)*. Saarbrücken: vdm Verlag Dr. Müller.

LB Barking and Dagenham (2010). *Planning for the future of Barking and Dagenham. Core strategy*. London: London Borough of Barking and Dagenham.

Lefebvre, H. (1971). *Everyday life in the modern world*. New York: Harper & Row.

Madanipour, A., Knierbein, S. and Degros, A. (Eds.) (2014). *Public space and the challenges of urban transformation in Europe*. London: Routledge.

Malik, K. (2018). We're now confusing diversity and equality. Which is our priority? *Guardian*, [Online] 17 June 2018. Available at: www.theguardian.com/commentisfree/2018/jun/17/we-are-confusing-diversity-and-equality-which-is-our-priority [Accessed 27 June 2018].

Manzi, T., Lucas, K., Lloyd-Jones, T. and Allen, J. (2010). *Social sustainability in urban areas*. Abingdon: Earthscan.

Mayor of London (2018). *London housing strategy*. London: Greater London Authority. Available at: www.london.gov.uk/sites/default/files/2018_lhs_london_housing_strategy.pdf [Accessed 9 Nov. 2108].

Mayor of London (2017). *The London plan. Spatial development strategy for Greater London. Draft for consultation*. [Online] London: Greater London Authority. Available at: www.london.gov.uk/sites/default/files/new_london_plan_december_2017.pdf [Accessed 7 Apr. 2018].

Mayor of London (2016). *Housing supplementary planning guidance*. [Online] London: Greater London Authority. Available at: www.london.gov.uk/sites/default/files/housing_spg_revised.pdf [Accessed 7 Apr. 2018].

Mayor of London (2012). *Shaping neighbourhoods. Play and informal recreation.* [Online] London: Greater London Authority. Available at: www.london.gov.uk/what-we-do/planning/implementing-london-plan/supplementary-planning-guidance/play-and-informal [Accessed 7 Apr. 2018].

Mayor of London (2007). *Planning for equality and diversity in London. Supplementary planning guidance to the London Plan.* [Online] London: Greater London Authority. Available at: www.london.gov.uk/what-we-do/planning/implementing-london-plan/supplementary-planning-guidance/planning-equality-and [Accessed 7 Apr. 2018].

Mayor of London (2004). *The London plan. Spatial development strategy for Greater London.* [Online] London: Greater London Authority. Available at: www.london.gov.uk/what-we-do/planning/london-plan/past-versions-and-alterations-london-plan/london-plan-2004 [Accessed 7 Apr. 2018].

Mitchell, Don (2003). *The right to the city. Social justice and the fight for public space.* New York: Guilford Press.

Mitchell, D. (1996). Introduction. Public space and the city. *Urban Geography*, 17(2), 127–131.

ÖROK – Geschäftsstelle der Österreichischen Raumordnungskonferenz (2011). *Austrian spatial development concept 2011.* [Online] Vienna: Eigenverlag. Available at: www.oerok.gv.at/raum-region/oesterreichisches-raumentwicklungskonzept/oerek-2011/download-oerek-2011.html [Accessed 9 May 2018].

Rathkolb, O., Autengruber, P., Nemec, B. and Wenninger, F. (2013). *Straßennamen Wiens seit 1860 als "Politische Erinnerungsorte" (Street names of Vienna since 1860 as "Political memorial places").* [Online] Vienna: Kulturabteilung der Stadt Wien. Available at: www.wien.gv.at/kultur/strassennamen/strassennamenpruefung.html [Accessed 9 May 2018].

Reinprecht, C., Dlabaja, C., Stoik, C., Kellner, J. and Kirsch-Soriano da Silva, K. (2015). *Forschungsbericht der Studie Besiedelungsmonitoring Seestadt Aspern (Research report of the study Settlement Monitoring Seestadt Aspern).* [Online]. Available at: www.wohnbauforschung.at/index.php?inc=download&id=5808 [Accessed 9 May 2018].

Sánchez de Madariaga, I. (2013). Mobility of care. Introducing new concepts in urban transport. In: I. Sánchez de Madariaga and M. Roberts (Eds.), *Fair shared cities. The impact of gender planning in Europe.* Farnham: Ashgate, 33–48.

Senatsverwaltung für Stadtentwicklung (2011). *Gender mainstreaming in urban development.* Berlin: Kulturbuch-Verlag GmbH.

Stadtentwicklung Wien – Magistratsabteilung 18 (MA 18), Stadtentwicklung und Stadtplanung (Urban Development Vienna – Municipal Department 18, Urban Development and Planning) (2017). *Städtebauliche Kennwerte – Analyse in Praxis und Modell mit Schwerpunkt Freiraumplanung (Manual of urban development indicators – Analysis of practice and models on the topic of open space planning).* Werkstattbericht, Nr. 167. Vienna.

Stadtentwicklung Wien – Magistratsabteilung 18 (MA 18), Stadtentwicklung und Stadtplanung (Urban Development Vienna – Municipal Department 18, Urban Development and Planning) (2005). *Stadtentwicklungsplan Wien 2005 – STEP 05 (City Development Concept 2005 – STEP 05).* [Online]. Available at: www.wien.gv.at/stadtentwicklung/strategien/step/step05/download/index.html [Accessed 9 May 2018].

Stadtentwicklung Wien – Magistratsabteilung 18 (MA 18), Stadtentwicklung und Stadtplanung (Urban Development Vienna – Municipal Department 18, Urban Development and Planning) (1994). *Stadtentwicklungsplan für Wien 1994. Beiträge zur Stadtforschung, Stadtentwicklung und Stadtgestaltung (City Development Plan for Vienna*

1994. Contribution to urban research, development and design) Bd. 53. [Online]. Available at: www.wien.gv.at/stadtentwicklung/studien/b006750.html [Accessed 9 May 2018].

Tovatt Architects & Planners and Projektteam Flugfeld Aspern (2007). *Masterplan Flugfeld Aspern (Masterplan airfield Aspern)* [Online]. Available at: www.wien.gv.at/stadtentwicklung/projekte/aspern-seestadt/planungsprozess/masterplan.html [Accessed 9 May 2018].

TURAS (2016). *Transitioning Towards Urban Resilience and Sustainability (TURAS) final presentation* [Online]. Available at: www.uel.ac.uk/sri/research-projects [Accessed 9 November 2018].

Vienna City Administration MA 18 – Municipal Department 18, Urban Development and Planning (2014). *STEP 2025. Urban Development Plan Vienna.* [Online]. Available at: www.wien.gv.at/stadtentwicklung/studien/pdf/b008379b.pdf [Accessed 9 May 2018].

Wien 3420 aspern development AG (2018). *Fortschreibung Masterplan. Stand der Planung 2017 (Update master plan. Status of planning 2017).* [Online]. Available at: www.aspern-seestadt.at/downloads/fortschreibung_masterplan [Accessed 9 May 2018].

Wiener Stadtgärten, Magistratsabteilung 42 (MA 42) and Stadtbaudirektion (Municipal Department Parks and Gardens (MA 42) and Executive Group for Construction and Technology (2005). *Planungsempfehlungen zur geschlechtssensiblen Gestaltung von öffentlichen Parkanlagen (Planning guidelines for gender-sensitive park design).* [Online] Vienna. Available at: www.wien.gv.at/stadtentwicklung/alltagundfrauen/pdf/planung.pdf [Accessed 9 May 2018].

Zibell, B. with Dahms, N.-S. and Karácsony, M. (2006). *Bedarfsgerechte Raumplanung. Gender Practice und Kriterien in der Raumplanung. Endbericht. Langfassung (Requirement-oriented spatial planning. Gender practice and criteria in spatial planning).* Materialien zur Raumplanung, Bd. 20. [Online] Salzburg: Land Salzburg. Available at: www.alpine-space.org/2000-2006/uploads/media/GENDERALP__ROSP_lang.pdf (German) www.gender-archland.uni-hannover.de/1212.html (English).

5 Gender sensitivity in neighbourhood planning

The example of case studies from Vienna and Zurich

Ulrike Sturm, Stephanie Tuggener, Doris Damyanovic and Eva Kail

Aims and content

The authors examine how gender-sensitive planning policies are visible at the neighbourhood level. The chapter focuses on a transnational comparison of relevant guidelines for neighbourhood planning and good practice examples for housing and public spaces in Austria and Switzerland. Both countries have different practices concerning gender sensitivity in urban planning and spatial development. The chapter introduces important guidelines for Vienna as the Austrian example and the Swiss "Housing evaluation system" (Wohnungs-Bewertungs-System (WBS)) and "Lares" criteria for Zurich as an example from Switzerland, both with respect to housing development and the planning of public spaces. After describing the framework for the neighbourhood level, the authors explore the underlying strategic objectives and urban development models of the guidelines and relevant criteria and indicators for housing development and public parks. The projects Nordbahnhof and Attemsgasse Ost in Vienna and Kalkbreite in Zurich are presented as good practices for housing development, and Rudolf-Bednar Park, Vienna, and Pfingstweidpark, Zurich, as good practices for planning public space. After each section, differences and similarities of the Austrian and Swiss case studies are discussed. To conclude, the authors summarise the lessons learnt from the comparative study and formulate some suggestions for future action.

Introduction – overview of the framework for neighbourhood planning in Austria and Switzerland

Due to the different legal planning structures (see Chapters 2 and 4), the framework for neighbourhood planning is described at a federal/city level for Vienna and at the national level for Zurich.[1]

The neighbourhood level framework in Austria (Vienna)

Gender-sensitive criteria in the context of the Vienna Housing Fund (Wiener Wohnfonds)

The Vienna Housing Fund is owned by the City. It is tasked with buying sites for subsidised housing projects, organising developer competitions for new subsidised housing projects that exceed 300 dwellings, making quality assessments for smaller new projects via a regular board and distributing subsidies for urban renewal (wohnfonds_wien, undated). Gender-sensitive planning criteria, developed by the Coordination Office, sit alongside the general quality criteria of the Vienna Housing Fund. Compared to other European cities, Vienna has a high percentage of social housing. The social housing system (initiated by the famous "Red Vienna" programme after the First World War) offers 220,000 dwelling units owned by the City and there are additionally 180,000 units of subsidised housing,[2] which combine to accommodate nearly 40 per cent of the Viennese population of 1.9 million (Vienna City Administration MA 23, 2017: 5). Ten years ago, around 80 per cent of newly built dwelling units were subsidised. Due to rapid growth and economic development making housing an attractive investment, this has decreased to around 50 per cent – still a high percentage. Therefore, using gender-sensitive criteria to assess subsidy applications impacts considerably on the construction of high-quality, affordable housing. Gender planning experts were sitting on the juries of developer competitions for over ten years. Since 1995 the Vienna Housing Fund criteria have been based on a three-pillar model consisting of planning, economy and ecology. Gender criteria provided an addition to the planning-quality pillar in the context of a gender-specific preliminary review. The model was supplemented in 2009 with social sustainability as a fourth pillar. Many gender+ criteria were integrated into the criteria list of the Fund (see Table 5.1; Damyanovic, Reinwald and Weikmann, 2013: 88–89).

Vienna's manual "Gender mainstreaming in urban planning and urban development"

The manual "Gender mainstreaming in urban planning and urban development" (GM Manual) provides a basis for a target-group-oriented assessment of projects, a set of indicators for different levels of planning, and functional guidelines and methods for gender-sensitive assessment and evaluation of projects. The instruments at the strategic level are the Urban Development Plan 2025 (Vienna City Administration MA 18, 2014), followed by sectoral concepts for green and open space, mobility and public spaces (Vienna City Administration MA 18, 2015a, 2015b; Stadtentwicklung Wien MA 18, 2018).Three planning activities influence neighbourhood plans: (1) Basic decision-making in urban development concepts and master plans, (2) The transfer of master plans and urban design proposals to legally binding land-use plans (Flächenwidmungsplan) and building control plans (Bebauungsplan), (3) Project planning for public spaces, residential and public service buildings. In the GM Manual, the key term "high-quality urban density" guides the discussion of urban planning. The provision of private or semi-public open spaces at the block or site level is considered. A sunlight indicator for these spaces provides an additional quality parameter. The manual states the acceptable number of storeys in sight or earshot of the street or the open space and defines the maximum block depths (depending on north-south or east-west orientation) for flats extending throughout the block – to ensure sufficient light in the innermost areas. These parameters, when combined with indicators for public green spaces from the guidelines for green and open spaces, provide differentiated measures of density. The GM Manual therefore recommends a meticulous quality check of the urban planning design from a net FAR[3] of 2.0 (Damyanovic, Reinwald and Weikmann, 2013).

Vienna's "Manual of urban development indicators"

Urban development indicators and guidance values are quantitative expressions of planning cultures and attitudes. In Vienna numerous indicators for different functions exist, many with different basic reference values. The intention of the "Manual of urban development indicators – analysis of practice and models for open space planning" (Stadtentwicklung Wien MA 18, 2017) was to harmonise these norms and parameters for urban design and to check for potential conflicts. The specification of urban planning indicators, including those in the GM Manual, are analysed and presented below for the most extensive urban development areas of Vienna, the Nordbahnhof, the Hauptbahnhof-Sonnwendviertel and Aspern Seestadt. The fact that these developments are well known aids understanding the guidance values. All three areas are "brownfield developments" previously held by one owner. The freedom for urban design was therefore great. Nevertheless, in all three areas, a looser/variable perimeter block style of development predominates.

The manual had another goal in the modelling of three-dimensional master plans. Building on the most significant normative indicators and empirical average values for the areas under analysis, three different models for blocks were developed, which were then combined in a second step into 50 ha model districts with perimeter blocks as the basic typological form. The application of the urban development indicators, in various forms, in a generalised model also helps with their visualisation. Through comparison, the interaction between the different indicators and parameters becomes clearer. The model can be used as the basis for an initial quality check of the relevant parameters at the beginning of the process or in the subsequent stages. For example, an important new indicator, m^2 of unbuilt space per inhabitant, was introduced instead of m^2 green space per inhabitant, thereby synthesising urban layout and open space parameters, whereas the degree of green space depends on the actual project.

Framework for the neighbourhood level in Switzerland (Zurich)

Not-for-profit housing in Zurich

Two-thirds of Swiss dwelling units are rented (Swiss Federal Statistical Office, undated a). In Zurich almost 75 per cent of the 425,000 inhabitants live in rented dwellings; another 15 per cent are members of housing cooperatives (Swiss Federal Statistical Office, undated b). Switzerland has a long tradition of not-for-profit housing. From the early twentieth century, Zurich housing programmes followed three objectives: predictive purchase of land, municipal housing construction and promotion of housing cooperatives, which offer affordable housing for specific user groups (e.g. employees) or members (Mehr als Wohnen, 2007: 18). The City of Zurich has the highest percentage of housing cooperatives, also owning a large share of the land of the cooperatives. Additionally, the City owns municipal dwelling units: approx. 6,500 in municipal housing estates and over 2,500 in individual properties (Stadt Zürich, 2016: 6). In 2011, the electorate of Zurich consented to a housing programme to increase the percentage of not-for-profit housing in the city to one-third; this has now almost been reached (Stadt Zürich, undated b).

Housing evaluation system (WBS), Switzerland

To define a quality standard for housing, in 1969 the Swiss Federal Research Commission on Housing Construction (Eidgenössische Forschungskommission Wohnungsbau) initiated an investigation into housing needs. Building on this, the "Housing evaluation system" (WBS) was commissioned as a "tool for the qualitative evaluation of residential building plans in the context of the law promoting residential building and home ownership". The first version of the WBS appeared in 1975 (Eidgenössische Forschungskommission Wohnungsbau, 1975). Since then, the WBS has undergone three revisions and updates in 1986, 2000 and 2015, commissioned by the Swiss Federal Office for Housing (Bundesamt für Wohnungswesen BWO; see BWO, 1979, 1986, 2000, 2015). The basic structure introduced in 1975 defines three "sets of objectives" to be applied under the headings of dwelling units, residential areas and locational infrastructure, with criteria and indicators for quality assessment. In the latest version the principal structure is preserved, with the three headings changed to (1) Location, (2) Residential area and (3) Dwelling unit. The "usefulness" (1975), or "utility" (2015) of the residence is evaluated with regard to all the three headings. In the WBS revisions, carried out by the Federal Ministry for Housing, the changing needs and requirements for housing design and construction were considered. The WBS was transformed over time from a tool for experts, originally used only for funding allocation, into a tool for general use and quality checking. WBS 2015 can be accessed online for the planning, assessment and comparison of residential buildings (BWO, 2015). Instead of the 75 criteria of 1975, there are now 25. In determining the so-called "utility" of a residence, specific uses and benefits for residents are of primary importance. Many WBS evaluation criteria have found their way more or less unaltered into the standards of the German Sustainable Building Council (Deutsche Gesellschaft für Nachhaltiges Bauen (DGNB), undated) and the Swiss Network of Sustainable Building (Netzwerk Nachhaltiges Bauen Schweiz (NNBS), 2016: 19).

The WBS formulates housing quality criteria which are committed to an everyday life approach; they do not, however, explicitly address gender issues. This is still the case in the latest version, although since the 1990s gender issues have been directly addressed in planning in Switzerland.[4] With regard to gender politics in general, EU laws were adopted in Switzerland (Amsterdam Treaty on Gender Mainstreaming 1999). Gender mainstreaming has been obligatory since 2004 for Swiss Federal Offices.

"Lares" – gender-sensitive and everyday-needs-specific planning and construction in Switzerland

"Lares" was conducted between 2006 and 2012 and initiated by the technical associations FFU – FachFrauen Umwelt; P, A, F. – Planung, Architektur, Frauen; SVIN – the Swiss Association of Female Engineers; the commission "Frau und SIA"; the Bern Urban Planning Office and the City of Bern Specialist Centre for Equality Between Women and Men. The Federal Office for Gender Equality funded the project. "Lares" had two main objectives (Lares, 2012): (1) To establish gender-sensitive and everyday-needs-specific construction and planning as an important dimension of sustainability in building and planning; (2) To balance the relationship between women and men in construction and planning decisions.[5] In six years "Lares" assessed more than 30 planning and building projects with regard to gender and the needs of corresponding user groups. The projects varied regarding size, planning level and subject (e.g. office buildings, higher education institutes, public spaces). "Lares" defined a standardised process to assess projects on the basis of criteria for gender sensitivity and everyday needs. In "Lares" projects, planning and construction were assisted by women specialists and checked against these criteria. An evaluation of "Lares" projects showed that the gender assessments contributed to the quality of the projects (Siegel, 2012).

The two main objectives of "Lares" were further supported by female specialists serving on competition juries and promoting the implementation of gender criteria in competition programmes and project developments. Besides specific spatial and architectural plans, "Lares" assessed the Swiss housing evaluation system WBS in 2008 from a gender perspective during its third revision. The focus on utility as a measure of quality and the fact that the WBS considers a wide diversity of housing concepts and forms of living were found to be positive. "Lares" suggested the implementation of structural and process-specific criteria in WBS (Lares, 2008), e.g. proposing that equal shares of men and women serve on committees as a structural criterion and recommended the participation of future women inhabitants or gender experts as an important process criterion. Participation was included as a general – not gender-related – process criterion in the WBS 2015. Overall, "Lares" identified the need for more research on gender and housing in the Swiss context.

Since 2012 and the end of funding, "Lares" has been organised as an association. The aims and duties of the association remained largely the same as during the "project period". However, the lack of a professional office made it more challenging to find project partners for gender assessments, so the number of assessments has declined since 2012. But "Lares" remains very active in providing information, organising exchange platforms and raising awareness.

Differences and similarities

Vienna as a federal state is responsible for the housing-subsidies law and the distribution of financial means, as a city it is responsible for neighbourhood planning. The City of Vienna has been addressing gender-sensitive planning for two decades. In 2001 the Co-ordination Office took over the central task of evolving and establishing the novel strategy of gender mainstreaming as a process-orientated strategy to safeguard quality in planning. But all planning departments of the City of Vienna contributed to the development of criteria on different planning levels as the basis for the Vienna GM Manual, elaborated in 2012 and published in 2013 (Damyanovic, Reinwald and Weikmann, 2013: 10–15), which also contributed to the 2025 Viennese Urban Development Concept.

In Switzerland, neighbourhood planning is undertaken at the city level, steered, however, by national and cantonal law (see Chapter 2). With the Swiss housing evaluation system WBS, the Swiss Federal Research Commission on Housing Construction created a national tool to secure the quality of residential complexes as early as 1975. This helped ensure the high quality of Swiss not-for-profit housing in general, even influencing the private sector. WBS was installed as a quality check for housing projects with an everyday-needs approach, and was constantly revised to meet current needs. Yet even in the latest 2015 version, gender criteria were not explicitly included, despite WBS being assessed from a gender perspective in 2008 (BWO, 2015). "Lares", first as a project, meanwhile as an association, fosters explicit gender discussion in the fields of public space and building, defining a standardised process of assessment for projects that explicitly combines gender-sensitive *and* everyday-needs criteria. Additionally, projects were assisted by women specialists and checked against these criteria, enhancing project quality. Despite these activities, gender-sensitive planning is not institutionalised in Switzerland.

Comparison of the guidelines for neighbourhood planning in Austria and Switzerland

Social and spatial planning specific strategic objectives and urban development models

In the Vienna GM Manual, the following strategic objectives of gender-sensitive planning are named:

A Taking into account and supporting the compatibility of family duties and paid work
B Equitable resource distribution through awareness of different spatial needs
C Attractiveness, safety and security of the housing environment
D Representation based on equitable participation and involvement of all groups in development and decision-making processes

The objective of "compatibility of family and professional life" (A) relates to the ability to combine productive and reproductive work. The issue of women's safety in public spaces, always much discussed in the gender debate in the context of planning, is often addressed as an objective together with the "attractiveness and safety of the residential environment" (C). The objectives of "equitable distribution of resources" (B) and "representation … of all groups" (D) ensure that population diversity is considered "in development and decision-making processes".

The GM Manual also lists the following mainstream urban development models as associated with gender-sensitive planning objectives:

E Strengthening a polycentric urban structure
F A city of short distances
G High-quality public space
H Promotion of environmentally friendly means of transport
I A safe city
J A barrier-free city
K Planning and construction geared to the requirements for daily life

WBS shows similarities with the planning models formulated in the Vienna Manual: "Deconcentrated concentration" was referred to as a guiding principle and urban development model in the first edition (Eidgenössische Forschungskommission Wohnungsbau, 1975: 38), corresponding to the Viennese "strengthening of a polycentric urban structure" (E). In view of the location, the "quality of infrastructure" is a central focus[6] in WBS and postulates proximities corresponding to the urban development model of a "city of short distances" (F), even if the Swiss WBS as a national instrument is not solely oriented to urban contexts. The new approach of WBS 2015 focuses on the "supplementary utility of the location" (BWO, 2015). This idea is based on the urban development model of "social mix and mixed uses" and the incorporation of new buildings into existing neighbourhoods to create mutual benefit. The Vienna Manual also addresses the question of suitability for various groups on a strategic level as "equitable distribution" for various demands on space (B). In WBS, mobility and open space are assessed under the objectives of good access for public and non-motorised transport. In the Vienna Manual, these issues are addressed as "environmentally friendly means of transport" (H). The principles of "high-quality public spaces" (G), "safe city" (I) and "barrier-free city" (J) are not named explicitly in WBS. Reference, however, is made at residential-area and dwelling-unit level to the building norm SIA 500 "barrier-free buildings" of the Swiss association of engineers and architects, which is obligatory for new buildings (SIA, 2009).

Specific gender criteria and indicators

One important feature of manuals is that they formulate specific criteria for urban design, together with measurable indicators. Both the Swiss WBS 2015 and the Vienna Housing Fund present indicator tables for inputting qualitative and quantitative descriptions. In WBS, these are directly linked to evaluation points, which, added together, give a numerical value for utility on all three levels and an overall utility value (BWO, 2015). For "City structure, space creating and quality of living" the Vienna GM Manual defines the following gender-relevant planning objectives (criteria) and formulates indicators (Damyanovic, Reinwald and Weikmann, 2013: 33–44), some of them measurable, some based on qualitative descriptions:

Urban structure

a Safeguarding framing conditions for high housing quality in the neighbourhood
b Tightly knit route network with appropriately sized building volumes
c Variety of public and open spaces

Development of local centres and provision of local shops and service providers

d Development of local centres
e Availability of easily reachable basic shopping outlets and service providers in the immediate vicinity of flats

Social infrastructure

f Systematic planning and inclusion of social infrastructure facilities for all population groups
g High-quality planning for kindergartens and schools with respect to their location, plot size, plot configuration and building height

Housing construction and open/green spaces for individual lots

h Range of flat typologies and flat layouts complying with various tenancy or ownership arrangements financing and cost requirements
i Stimulating good neighbourly relations
j Adequately dimensioned private and communal open spaces for everyday tasks, children's play and leisure

Increased objective and subjectively perceived security and safety

k Effective and clear-cut spatial orientation and social control in the neighbourhood

For "public spaces and mobility", the following planning objectives (criteria) are listed together with measurable indicators (Damyanovic, Reinwald and Weikmann, 2013: 45–49):

Public open spaces

l Differentiated design of public open spaces in the urban quarter
m User-friendly distribution of public open spaces in the urban quarter

Mobility

n Promotion of pedestrian and bicycle traffic and of public transport by a fair division of street space
o User-friendly and route-optimised public transport.
p User-friendly route networks for pedestrians and cyclists

The Vienna GM Manual also includes several issue-specific criteria and indicator lists for different planning levels concerning everyday life and gender-fair urban planning concepts and master plans (Damyanovic, Reinwald and Weikmann, 2013: 58–59), area and development plans (67–68), public spaces (82–84)[7] and residential or commercial buildings (88–89).

Compared to WBS – and to the Vienna Housing Fund – the Vienna GM Manual goes into more detail on the topics of the "development of local centres and provision with local shops and service providers", "increased objective and subjectively perceived safety and security" and "public open spaces". This reflects the fact that the Vienna GM Manual focuses more on higher planning levels and considers in particular district planning. "Gender mainstreaming as a comprehensive planning strategy" is allocated a chapter of its own for planning processes. WBS concentrates on the evaluation of housing itself and is therefore less expansive in the urban context but significantly more detailed at the residential-area and dwelling-unit levels. A corresponding level of detail in these matters can be found in the assessment criteria for eligibility for housing subsidies according to the Vienna Housing Fund, which are not fully explored here. For a better overview, some key criteria and relevant indicators, focusing on the neighbourhood level as the main topic of this chapter, are shown in Table 5.1.

Table 5.1 Comparison of the content criteria of the Vienna GM Manual, Vienna Housing Fund and WBS

	Criteria	*Vienna GM Manual*	*Vienna Housing Fund*	*WBS CH*
a	Ensuring conditions for high housing quality in the neighbourhood.	High housing quality is very likely to be secured with a FAR of up to 2.0.	Integration into the urban context concerning its scale etc.	The urban design of the residential complex (size, typology) fits well with its context (or is a positive contrast).
c	Differentiated availability of public squares and open spaces.	Public open spaces serve different functions taking account of varying interests and purposes (leisure, play, sports, etc.).	Open space can be appropriated by different user and age groups, providing for noisy activities and robust materials.	Open space can be appropriated by different user and age groups, including covered spaces.
f	Systematic planning and inclusion of social infrastructure facilities for all population groups.	Spatial preconditions (e.g. multi-purpose rooms or ground-floor premises with higher ceilings) that favour the establishment of social infrastructure and service facilities.		Work spaces are provided. The residential complex has social institutions (kindergartens, care for the elderly, etc.).
h	Wide range of dwelling unit typologies and layouts complying with various tenancy or ownership structures, financing and cost requirements.	Small two- and three-room flats with separately accessible rooms, etc. Adaptability of flats is e.g. supported by demountable partitions, rooms of equal size or more than one entrance to rooms (separate entrances).	Social mix by different dwelling forms. Opportunities for different (residential) cultures and different housing.	Potential to enhance diversity in the neighbourhood; wide range of dwelling-unit layouts. At least two movable partitions or possible wall openings per flat.

continued

Table 5.1 Continued

	Criteria	*Vienna GM Manual*	*Vienna Housing Fund*	*WBS CH*
i	Stimulating good neighbourly relations.	Involving residents at an early date is recommended. The share of communal and secondary rooms (playrooms, communal lounges, laundry rooms, storerooms, basement compartments, etc.) should be at least 3–5 per cent of the total floor area of the flats.	Concepts of co-determination in planning, building and use. Management in the case of self-build (cohousing projects).	Stakeholders are identified and participate regularly during the planning process. 2 m^2 of common space provided per inhabitant.
j	Adequate provision for private and communal open spaces for everyday tasks, children's play and leisure.	The open spaces of the project are designed in a gender-sensitive manner and take account of different patterns of space appropriation.		Open space provisions in addition to hard courtyards: green spaces such as meadows, bushes and trees.
m	User-friendly distribution of public open spaces in the urban quarter's open spaces.	The individual open spaces are networked by squares, pedestrian zones, greened streets or streets with adjacent green zones. 3–5 m^2 of public space (parks, squares, etc.) per inhabitant of a housing project is recommended.		Open space is well connected to the surrounding neighbourhoods.

n	Promotion of pedestrian and bicycle traffic and of public transport by fairly dividing street space.	There are sufficient collective garages to reduce the number of cars parked on the street. The vacated space serves to improve conditions for non-motorised traffic and public transport. Mobility management offers alternatives to private car use. One bicycle parking slot is provided for every 30 m^2 of dwelling space. Parking spaces for prams are likewise provided.		Underground, barrier-free car parking with natural lighting. Car sharing is available. Sufficient storerooms for prams and bicycle parking are provided.
o	Public transport is user-friendly and route-optimised.	The public transport stops can be reached on foot and without physical barriers within a distance of 500 m (Underground) and 300 m (tram and bus).		Max. 1 km from centre of the residential area to the nearest public transport stop.
p	User-friendly route networks for pedestrians and cyclists.	A close-knit, walkable and barrier-free route network with adequate atmospheric quality is in place.		The mobility concept of the residential area creates new networks for pedestrian and bicycle traffic.

Assessment of best practice – "site planning"

Case study Vienna Nordbahnhof Site "Block J"

The Nordbahnhof area in the 2nd district of Vienna with 85 ha is one of the largest inner-city development zones. It is located between the city centre and the Danube, which is only two blocks away, surrounded by a neighbourhood dominated by late nineteenth century perimeter blocks and buildings, Gründerzeit. The Nordbahnhof is a former railway station and thus a brownfield development. Half of the development has already been completed; the other half is scheduled for 2025. In 2025 10,000 new homes and 20,000 new workplaces will be provided. The first planning mission statement dates back to 1994. The objective was to create an area of mixed-use perimeter blocks in several phases, first a band of offices, followed by residential buildings, the central Rudolf-Bednar Park (see below) and an educational campus, primary school and kindergarten. Although omitted in the first phase, a lot of gender-planning activities took place in later planning phases:

- Gender pilot projects like the competition for the campus and Rudolf-Bednar Park
- A gender-planning assessment as part of the prequalification for the two-phase urban design competition of the last section
- Participation of a gender-planning expert in the juries of the two developer competitions

In the "Manual of urban development indicators" (see above) the first phase of residential buildings of Nordbahnhof was analysed. A workshop with developers, architects, landscape architects and planners from the zoning department discussed favourable factors for usability for all, the attractiveness and openness to the surroundings, and focused on good block dimensions and the quality of open spaces within the blocks. Each participant could vote and argue for one block of the three analysed areas from the manual. Block J received (with Block D9 in Aspern Seestadt with five cohousing projects) the highest number of votes.

The block was developed from 2009 to 2013, primarily as an outcome of a thematic developer competition on "Intercultural Housing" in 2010. The competition area consisted of two sites for two non-profit developers, one offering 101 dwelling units, including 35 units for accredited refugees with a consultant's office for one of the most engaged NGOs (Integration house: Interkulturelles Wohnen Nordbahnhof). The other non-profit development site has two buildings, designed by different architectural offices: 48 low-rent dwelling units for migrants and young people (Wohnen mit scharf!), and an ambitious cohousing project with 40 dwelling units (Wohnen mit uns! – Wohnprojekt Wien). All three buildings offer a variety of common rooms and facilities. The block also includes 108 privately financed dwelling units in three buildings of the project

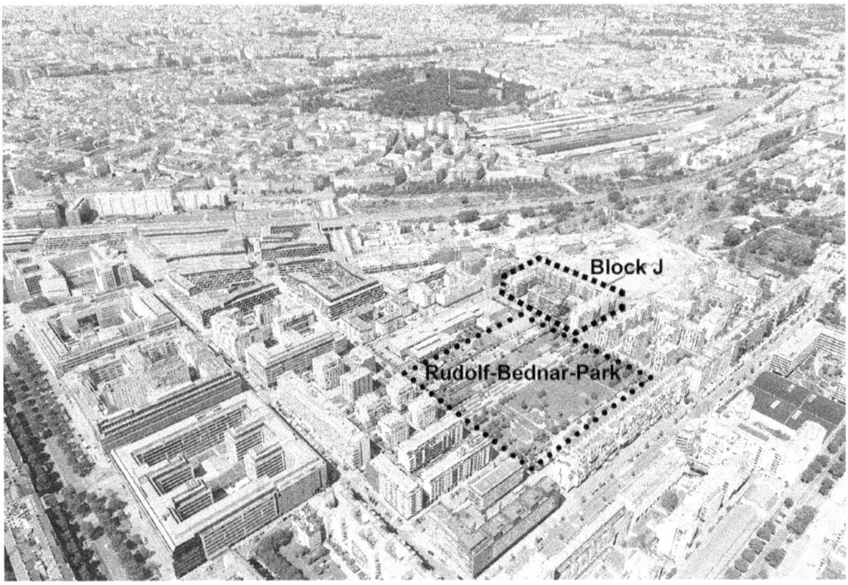

Figure 5.1 Aerial view of Nordbahnhof.

Source: © City of Vienna, City of Vienna, MA 18/C. Fürthner.

"New Park living" with poor infrastructure (just one common room and one doctor's surgery). This block allows a comparison between the privately financed development and the subsidised housing system with its high-quality approach including consideration of gender issues. In total, there are 297 dwelling units in the perimeter block with around 530 residents.

The block has a surface area of approximately 10,300 m^2 in a compact plot of 1:1.5 (width: length ratio) and a gross floor space of 32,500 m^2. The seven-storey development forms an open perimeter. The plot layout and size (123 m × 83 m) offers flexibility and would also permit a different urban design approach with different building typologies. The net FAR is 3.2. The unbuilt area (5,650 m^2) is divided into the compact inner courtyard (3,500 m^2) and five micro-areas. The useable open space per inhabitant is 4.7 m^2; the general open space ratio is 8.7 m^2 per inhabitant. The height proportion of the courtyard area is approx. 4:1 longitudinally and approx. 2:1 transversally.

Despite its location next to Bednar Park, the shared inner courtyard has an important function for residents. Children in particular appreciate it as a retreat zone, but it is also used for house parties and as a neighbourhood garden. The ratio of play areas and functionally unrestricted open spaces is well balanced. Users do not feel closed in or observed. In particular, the open lawn offers many possibilities. The junctions between the individual courtyards do not include any major changes in use or choice of materials. The cohesive design of the courtyard supports its communal nature. The design of the open space in the privately financed project, however, is that of an unambitious mainstream project, enclosed by a fence. With this exception the block is open to three sides with graduated levels of openness to the public. Also the 4m-high ground floor with its carefully designed façade has a positive influence on the park side. In particular, the cohousing project "Wohnprojekt Wien" with its "sociocracy", a collective social model for planning decisions and self-administration, won many prizes for sustainability and housing quality, and has had a positive overall effect on the block. The project also offers many neighbourhood facilities like a community-run coffee shop, a neighbourhood garden and in the underground storey a large well-lit rentable assembly room with a separate entrance. This was possible because of the low number of parking spaces, including two parking spaces for private car sharing.

The urban design for the whole area and of Site J was finalised prior to the publication of the Vienna manuals, although the gender-sensitive criteria for housing was already in existence. The jury for the developer competition recommended that the courtyard should be developed as a common open space with a consistent design. The winning projects "Interkulturelles Wohnen" and "Wohnen mit scharf!" were chosen for their high social contribution, their flexibility, the variety of their dwelling-unit typologies and for their social mix and affordable rents.[8]

Case study Vienna Attemsgasse Ost – urban design framework plan

The direct influence of the gender planning manuals on urban design can be shown in a more recent project, the urban design framework plan for the area Attemsgasse Ost. A new gender pilot project "DIE Stadt" was announced in the 2015 coalition agreement between the Social Democratic Party and the Green Party. The planning process started in summer 2017 and was finished in March 2018. The urban design framework for three hectares with around 600 dwelling units, including the enlargement of a small neighbourhood park, had to be developed as a basis for the land-use plan. As a first step, an urban design framework was produced by an external planning team of female architects, landscape planners and a traffic planner, all of them women, accompanied by district planning and land-use municipal department 21, and the City's gender-planning expert. This was combined with a gender-sensitive participation process and a gender advisory board, which discussed the different design proposals. As a first step, the prerequisites from a gender planning view of the area were intensively discussed. This was followed by a thorough discussion of the proposals of three different building typologies combined with different park layouts considering the quality of the open spaces, either for the public or within the sites. Green space was taken seriously as the backbone of development. In spring 2018 the dimension, shape and location of the new part of the park and the three sites were fixed, but the framework still included two design proposals for linear and courtyard layouts.

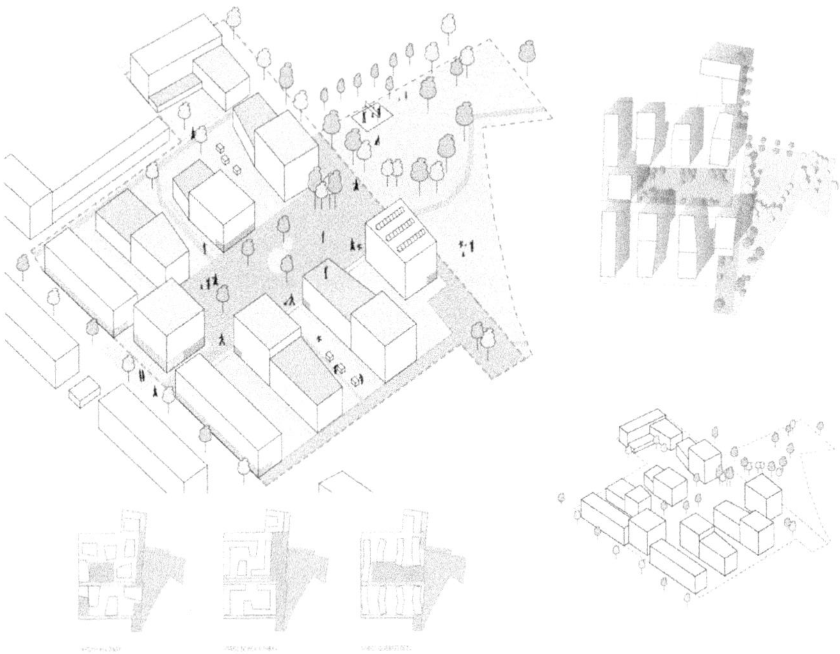

Figure 5.2 Masterplan for Attemsgasse from einszueins architektur ZT GmbH.
Source: © City of Vienna/MA 21.

The decision on the building typology and the general layout of the park enlargement was followed by the fine-tuning of the design proposals which took into consideration the neighbours' wishes as articulated in the participation process. The discussion of the board differentiated between the statements of the different groups. The fulfilment of most of the GM Manual criteria can be seen in the outcome. The potential shading of the public open space was an important criterion for the whole urban design. After debate, the dimension of the wing of the building on the main road in the north was reduced due to the manual's recommendation of 12 m in a north-south direction thus avoiding dwelling units that were only oriented towards the north and a busy street. In the linear proposal there is now only one main entrance to the new part of the park, one row of buildings was shortened to create a wider opening to the park. Half of the area is owned by Wohnfonds, and a gender-sensitive developer competition will be mounted in 2020. This project will also be part of the International Building Exhibition 2022 (IBA) in Vienna. The other area was bought by a private developer; here the building regulation plan will set a fairly tight framework, because there is no subsequent quality procedure.

Case study Zurich Kalkbreite

The Kalkbreite development in Zurich was built between 2012 and 2014 (Genossenschaft Kalkbreite, undated). The housing estate is leased and is located on the roof of a tram depot at Kalkbreitestrasse 6 in Zurich's city centre. The project was and is managed by the innovative housing cooperation Genossenschaft Kalkbreite, founded in June 2007. Under the City of Zurich's building laws, Genossenschaft Kalkbreite developed their first construction project together with the city, the transport association for Zurich and other partners in a broad participatory process. Participation is deeply rooted in the tradition of Swiss housing cooperatives; Genossenschaft Kalkbreite, however, made outstanding efforts to involve different target groups in the planning process and in the operation phase. The seven-storey building project, designed by Müller Sigrist Architekten AG, Zurich, was selected in an architecture competition in 2008. The building block with an open courtyard on the tram depot roof has a surface area of approximately 6,700 m^2 in a compact plot. The seven-storey development forms an open perimeter. The net FAR is 2.8 without the tram depot, 3.7 including the depot. Open space is divided into the compact inner courtyard (2,500 m^2) and additional rooftop areas. The useable open space per inhabitant in the courtyard is 10.9 m^2. The proportion (height to width) of the courtyard is roughly 2:1 in all directions (due to the irregular shape the number is averaged). There are many different types of dwelling units, 88 plus 9 "living jokers" (approximately 7,800 m^2 plus 630 m^2 common areas) for 251 inhabitants. The building includes innovative forms of living such as cluster flats, shared flats with a canteen kitchen, commercial uses and services mainly at street level, and a variety of social and cultural institutions, including a cinema (approximately 4,780 m^2).

The objectives of the development were formulated by Genossenschaft Kalkbreite in 2008, stating that the estate was to meet high standards with respect to social, economic and ecological requirements. Criteria were defined in accordance with the guidelines of the Swiss associations of engineers and architects SIA "Standard guide to SIA norm 112 – Sustainable construction – Structural engineering" (SIA, 2005) and "Leaflet 2050 and documentation 0246. Sustainable spatial development – Local and regional planning" (SIA, 2015).[9] To ensure sustainability the project was monitored in its planning phase according to the stipulated criteria (durable Planung und Beratung GmbH and raumdaten GmbH, 2013). Additionally, the final construction project was evaluated under WBS 2015.[10] With regard to its location, the building project of Kalkbreite received the highest rankings for its mix of dwelling-unit layouts and different uses (living, commercial, social and cultural). The project is highly interconnected with the surrounding district and well connected to public transport. It evolved from a carefully designed participatory process within the housing cooperative. There is a good supply of open areas for different uses. The residential complex was rated highly for traffic management such as car sharing and speed reduction, sufficient spaces for common activities and storage areas instead of large private rooms as well as well-designed entry foyers for encounters and communication. As inhabitants contractually agreed not to have their own cars, the provision of car parking is low, which was officially approved by the City of Zurich. This resulted in a low rating in WBS. With regard to the residential standards as defined by WBS, Kalkbreite performs well in terms of dimension and the functionality of rooms and kitchens. It receives average ratings for provision with private exterior space, bathrooms or private storage. When formulating its goals and the criteria for pre-occupation monitoring (see above), the housing cooperative explicitly aimed at innovative layouts for new forms of households and at reaching higher sufficiency in the use of space. This explains the deviation from WBS standards, which were defined for ensuring high standards, not for particularly innovative layouts.

In general, it can be concluded that everyday living requirements on the level of the location and residential area were ideally met by the project. The level of participation is particularly striking. The development process was accompanied by nine working groups of housing cooperative members. The formulated goals and monitoring criteria do not explicitly introduce concepts of diversity or gender. Some aspects of a diverse social mix such as different incomes or different household types resulting from a variety of lifestyles and living conditions (singles and couples of different ages, single parents with children, families, etc.) are, however, mentioned.

In the post-occupancy phase Genossenschaft Kalkbreite was researched as a case study investigating the quality of neighbourhoods in housing cooperatives (Emmenegger, Fanghänel and Müller, 2017). The neighbourhood is defined as the interplay between everyday living qualities, organisational structures and built environment.[11] Considering "dedication and motivation", "space of possibilities", "possibilities for appropriation", "openness and privacy", "identification and encounters", "integration" and "organisational structures", the housing estate was, in general, judged as allowing for a variety of lifestyles, supported by a semi-institutionalised system of solidarity. It was apparent, however, that facilities are used by all, but with different levels of engagement. Specific groups – which can roughly be described as the "alternative Swiss middle class" – are overrepresented in the decision-making bodies. As no specific focus was laid upon gender issues, there was no explicit analysis of how activity patterns mirror gender roles. Nonetheless, even the highly participative approach of Genossenschaft Kalkbreite does not lead to an even distribution of target groups in decision-making. Gender roles – encompassing other intersectionality features – may be reiterated by the organisational structures of a Swiss housing cooperative, based on the values of specific user groups such as "alternative Swiss middle-class" members. The collected data should be analysed in a second investigation to shed light upon this point.

Differences and similarities

The sites in Vienna Nordbahnhof and Zurich Kalkbreite are of different character. Although both projects are brownfield developments, their locations and

Figure 5.3 Kalkbreite – view from the street 2017.
Source: © Schopp.

Figure 5.4 Kalkbreite – inner courtyard 2017.
Source: © Schopp.

site contexts differ. Site J of the Vienna Nordbahnhof area is part of a homogeneous urban development area of regular rectangular blocks and is attractively situated beside a broad car-free street and the new 3 ha Rudolf-Bednar Park, whereas Kalkbreite has a difficult triangular site in an unusual location in a hybrid neighbourhood, bounded by two busy streets and a lowered train track. Kalkbreite had to integrate a tram depot and with its mixed use provides almost the same number of workplaces as inhabitants, using the floors adjacent to the tram depot. Site J is almost only for residential use, although it still houses more offices and facilities than the blocks around. Both sites have well-dimensioned central courtyards and a convincing layout. Site J has (except for the private section) an inviting, open character. The inner courtyard of Kalkbreite is also open to the public, yet it became too attractive to outside visitors, despite its elevated position. As the courtyard is the major open space for residents (the private open spaces are limited to the roof and ramps), unrestricted public use became annoying. Apart from the courtyard, Kalkbreite has a lot to offer the neighbourhood in the building itself. In Site J and Kalkbreite the intention was to reach a high density (reflected in the FAR of over 3). From a gender perspective, the target of affordable housing often conflicts with providing a satisfactory amount of unbuilt, open space on the site. The designs of Site J and Kalkbreite have resolved this conflict in a remarkably positive way.

There are differences between the developers, and the development and planning processes. In Zurich, the whole development was an outstanding, bottom-up process by a very innovative housing cooperative with high ecological and social aims. The landowner is the City, which leased the land to the cooperative. With Genossenschaft Kalkbreite the project did not follow the normal mainstream, developer's logic. The layout, the architecture of the buildings and the internal layouts of the housing units and workspaces were developed within one single competition, but with great participation and even collaboration from the Genossenschaft. In the case of Nordbahnhof, the former railway area was initially owned by the national train company ÖBB. The long planning process had several phases with several actors. The first master plan represented ambitious mainstream planning, with its 3 ha central neighbourhood park Rudolf-Bednar, bordered by buildings (an education campus and residential buildings) and two pedestrian streets. The area is dominated by a grid of rectangular blocks; the land-use plan allows flexible perimeter structures of good dimensions (length, width, height). A developer competition for subsidised housing followed for the majority of the block area, with a procedure that already incorporates gender to a certain degree.

In both site developments, communal interests were given a higher importance than in many other development projects. In both cases, quality criteria were applied. Genossenschaft Kalkbreite developed their own criteria for the competition, based on Swiss standards. For the non-profit part of Block J in Vienna, the process was multi-layered. The quite flexible land-use plan was based on the urban design master plan and provided the fixed framework for the developer competition and the basis for the private development. For the competition the general criteria of the Wohnfonds applied and the gender expert in the jury used gender-specific criteria for the housing assessment. Specific demands were also formulated for this competition, with regard to the requirement for intercultural housing. This was again the basis for the cohousing project Wohnprojekt, where further quality criteria and approaches were developed during the social process, but not so well documented as for Kalkbreite. Both block developments are examples of outstanding mainstream development activities, Kalkbreite as a whole, "Block J" in some parts (with some gender aspects already integrated in the mainstream process of the developer competition). Both planning processes led to a high level of social and ecological project-specific standards and outstanding qualities, nestling under a gender-sensitive approach or quality assessment standards. In both cases the innovative outcomes are striking, also from a gender+ perspective. The transformative potentials might have been even higher with an explicit gender perspective, e.g. in the case of the intercultural projects of "Block J", as traditional gender roles are often strong in many of the migrant groups involved.

Physical structures cannot change gender roles, but can support functional demands like those of caregivers, and participation processes can have a transformative influence on gender roles. Therefore, a corresponding evaluation of the participation and decision-making processes would be very interesting.

In the more recent case of Attemsgasse the development is at an early stage. It involved decidedly gender-sensitive planning and a participation process based on the GM Manual. The annexe to the framework plan systematically presented the contribution of the design proposal towards the strategic goals and the relevant goals for the urban design scale as set out in the GM Manual.

Criteria for gender sensitivity in park planning

The bases for the following comparison of criteria for gender-sensitive park planning are, on one hand, the planning recommendations for the gender-sensitive design of public parks by the City of Vienna (included in the GM Manual) and, on the other hand, the criteria for the tendering process for Pfingstweidpark, Zurich, developed by two "Lares" experts. The Viennese planning recommendation was developed in a lengthy quality assurance process started in 1997, when a study found that girls aged 10–13 tended to withdraw entirely from parks and open spaces. Several pilot action and working groups initiated by the Coordination Office were the basis for the recommendation, made available for all contracts of Vienna's Municipal Department Parks and Gardens (MA 42) (Irschik and Kail, 2013). The criteria developed by the "Lares" experts built on their research, experience and exchanges within the "Lares" network. At the time the "Lares" experts were also in communication with the Coordination Office in Vienna.

The fundamental difference between the two sets of criteria is that in Vienna the guidelines are a general framework for public park design. The tender for

Rudolf-Bednar Park was site specific. In the Swiss case, the criteria were developed just for this one tendering process. In Zurich there are no gender-sensitive criteria regarding planning of public parks.[12] By comparing the content criteria of the two cases, similarities have been identified in four areas: spatial structure, safety and security, places for recreation and encounter and the concepts of the activity range for different groups/mix of uses.

Spatial structure

- Well-connected open spaces: open borders that permit visibility/views, good accessibility
- Differentiated spatial concept: structuring into sub-zones, variety of different spaces (functionalised and open-ended zones), minimising conflicting use of space, maximising synergies (e.g. bench next to playground), sheltered areas/privacy
- Good network of paths: variety of differentiated routes and paths, minimal width of main paths, circular route option, barrier-free paths
- Areas with potential for change, flexibility

Safety/security

- Illumination, good lighting for paths
- Clarity, orientation: good visibility and legible organisation of footpath system
- Good visibility, social control: visual axes, clearly designated park entrance zones, fostering frequent use, especially of the main paths

Places for recreation and encounter

- Covered areas, seating and lounging zones for parents (e.g. next to playground)
- Access to toilets
- Drinking fountain
- Provision of protected and sheltered areas for elderly people

Activity range/mix of uses

- Spatial and play-related offers: varied options, multiple offers (e.g. two tables for table tennis), integrative and multifunctional play equipment, possibilities for appropriation, high potential range of experiences (e.g. variety of plants)
- Sheltered areas/privacy: visible and non-visible parts, areas for rest, recreation and communication next to areas for ball games and sports
- Furniture: multiple pieces of furniture that are attractive for various user groups, seating

One particular difference is that the Viennese situation demands more explicit consideration of girls' needs, because at a certain age they disappear from public spaces. Furthermore, in recent years in Vienna there has been increasing focus on the diverse needs of elderly people and ways of encouraging them to active lifestyles. The guideline also includes criteria for the whole planning procedure – like gender-sensitive participation, gender-sensitive work on site after construction (mobile youth work, park monitoring, etc.) and also additional special indoor meeting points for girls and boys close to the park. "Lares" promotes planning processes where gender should be taken into account in every planning phase, ideally from the very beginning (Lares, 2012). Pfingstweidpark served as a showcase here. To summarise, both approaches include a process-orientated approach for the integration of gender sensitivity in all phases of the planning process.

Assessment of best practice – "public space – parks"

Case study Vienna Bednar Park

Rudolf-Bednar Park (31,000 m²) is part of the urban development area "Nordbahnhof" (discussed above). In 2005, an invitation to tender was issued for the area by the Wiener Stadtgärten (Municipal Department Parks and Gardens, MA 42) in an international EU-supported park planning competition. The planning recommendations for gender-sensitive park design were an integral part of the tendering process. The preparations for the competition also comprised several workshops by different municipal departments (e.g. Architecture and Urban Design, Municipal Department Parks and Gardens, MA 42) as well as other relevant actors (municipal districts, the association "Wiener Jugendzentrum", the organisation "Parkbetreuung" offering park activities for children and young people, etc.) (Damyanovic, Reinwald and Weikmann, 2013: 85).

Among the projects submitted by landscape architects from Austria, Switzerland and the Netherlands, that of Hager Landschaftsarchitektur Zurich was unanimously selected as the winning project (Hager, Altermatt, and Posset, 2008). The first sod was turned in July 2007 and the park opened in September 2008. The spatial structure is based on a differentiated zoning concept with tranquil zones, sport zones and areas for play and exercise for all age groups. One essential feature is a cohesive curtain of 280 trees. The curtain of trees is oriented towards the nearby Danube and the tracks of the former train station. To the south of the park is an area for young people to skate and play streetball close to the school and kindergarten complex. The surfaced area in front of the residential complex "Wohnen am Park" is reserved for quieter uses. To the east of Rudolf-Bednar Park, there are shaped flowering shrubs in the lawn, referred to as "district gardens". These offer a peaceful area for all park visitors. The most striking feature are the many orange metal posts distributed over the entire park, marking children's play areas and supporting play equipment such as climbing ropes, swings, hammocks, etc., designed to suit all generations. There is a clearly

Figure 5.5 Bednar Park 2018.
Source: © Damyanovic.

structured footpath network with visual axes. Efficient illumination supports subjective feelings of safety (Stadtentwicklung Wien MA 18, 2013).

Overall, from a gender perspective the planning process for the design of Rudolf-Bednar Park has been a success in many respects. Due to its status as a "Gender mainstreaming pilot project" an analysis was made of the gender-sensitive open-space planning, an approach that had been challenging in terms of content for all involved and that heightened awareness of the subject. Rudolf-Bednar Park provides an open space where various play and exercise interests, places of recreation and encounter for communication or retreat, and safety needs are taken into consideration. It has become apparent that the process-orientated approach of gender mainstreaming importantly contributes to ensuring high quality in planning procedures. In the course of revising and finalising the details, the great potential of the competition entry was maximised with the assistance of gender experts and possible utility by varied groups further optimised in the interests of gender-sensitive design. So, e.g., following the gender guidelines the footpath network was expanded to allow circular walking routes in the park.

Case study Zurich Pfingstweidpark

Pfingstweidpark is a public park in Zurich-West, a former industrial area, which has been in a process of gradual transformation into a mixed-use area since the

1990s. During the course of this restructuring, new infrastructure such as schools and open spaces became necessary (Stadt Zürich, undated a). In 2010, the City of Zurich issued an invitation to tender for a park of about 12,000 m^2. In a second phase, a primary school was to be built next to the park. The competition conditions required a green district park that promoted a local identity for people living and working in the surroundings and considered their diverse need for open spaces and recreation (Grün Stadt Zürich, 2010). Complying with the wishes of the public and landowners, who were consulted in a participatory workshop in 2009, the City of Zurich attached importance to social sustainability and to the incorporation of various user needs during the process as well as in project evaluation.

In order to achieve these goals, female "Lares" specialists were involved in drawing up the competition programme, examination of preliminary projects and judging. For the competition programme, the "Lares" experts developed planning recommendations on the gender-sensitive design of public parks. As the process continued, they were available for consultation and provided support for the winning project during the pre-project phase (Hoppe, 2017). All teams were committed to drawing up a "social sustainability – gender mainstreaming plan", thus ensuring that gender concerns were taken into account in the competition. Furthermore, the "Lares" experts contributed the perspective of "requirements of use" in the judging. The winning project by Antón & Ghiggi landscape architects was judged to be of high quality regarding social sustainability and everyday needs. The park was opened in September 2015. Hoppe (2017) makes a positive assessment of the result.[13]

The Pfingstweidpark fulfils most of the criteria relevant for the design of a gender-sensitive park. The spatial concept is varied. Different kinds of functional and open-use zones are provided. The park is flexible and provides a wide variety of uses. Areas sheltered by trees, open areas for ball games and different seating (benches, moveable chairs) make it comfortable for various user groups. The arrangement of the pathways is logical and clear, facilitating orientation. The design uses visual axes to indicate where the nearest large street is, enhancing subjective feelings of safety. Most of the paths are barrier-free and allow circular routes. With varied offerings for play, leisure or congregating, the park takes the different needs of girls and boys and other user groups into account. Public toilets and a drinking fountain are provided.

One drawback is that the planned café was not built because of a lack of finance. A café would have enhanced the park's social function as a meeting point for social interaction. The explicit incorporation of gender criteria in the planning competition was unprecedented in Switzerland. This new approach made a "social sustainability – gender mainstreaming" plan an integral, compulsory part of the competition project and was introduced by the "Lares" specialists. Regrettably, this instrument has not become standard. Moreover, it is remarkable that gender experts were part of the jury. This emphasis on social sustainability and gender made a difference and contributed to the high quality of the park.

Figure 5.6 Pfingstweidpark 2018.
Source: © Tuggener.

Differences and similarities

Cities have direct ownership of their parks, which enables them to set specific thematic agendas like gender-sensitive park and playground design. In Zurich, this gender-specific approach was realised in one competition with the external gender expertise of "Lares" specialists. Although the outcome was convincing and the "Lares" experts had formulated stringent, easily transferable criteria, it remained a single pilot project. The authors assume that the reason for this is the low political value given to gender planning approaches in the mainstream system. This is underlined by the "Lares" experts' impression that the whole process was an "alibi" for the City of Zurich.

The gender-sensitive design of the Viennese Rudolf-Bednar Park was the result of a more consistent gender-mainstreaming process: six pilot projects and an evaluation of the design and user perspectives of several different park projects were used to formulate general gender-specific guidelines, which are part of the general park mission statement of the Municipal Department Parks and Gardens (MA 42). This is a mainstreamed basis for every new design or redesign of parks and playgrounds. The planning of Bednar Park was one of the lead projects that the planning and traffic departments had to realise in this period (see

Chapter 4). In this specific case of a new urban development area, an EU-wide competition was carried out by district planning and land-use municipal department 21 (in cooperation with MA 42), supported by the EU. Thanks to the existence of the Coordination Office, it was possible to provide the whole planning process with internal gender expertise.

Discussion and conclusion – gender sensitivity in neighbourhood planning

Neighbourhoods supplying different types of dwelling units with good quality (in terms of being affordable and useable), green, social and cultural infrastructure and good mobility conditions provide favourable prerequisites for well-functioning social relations and crucially influence the everyday life of residents. The importance of each of these factors, however, differs with people's life phases, social roles and cultural backgrounds. As the case studies showed, urban design and layout are highly gender-relevant issues for sustainable fair-shared cities.

Two principles of orientation: gender-sensitive planning or an everyday life approach

The gender concepts of the case studies stem from two different orientations concerning concept and strategic level of housing development: to formulate gender sensitivity explicitly and, on the other hand, to focus on an everyday life approach. In park design, however, there is an explicitly gender-oriented approach in the case studies of both countries with the application of gender criteria for the process and design criteria for parks. The comparison of the framework – different guidelines in the two countries – shows that a gender approach and an everyday life approach to housing development reach similar design criteria and indicators for the neighbourhood level. Their application to examples of building blocks shows that the existing manuals with an everyday approach (WBS) are supportive for (practical) needs (e.g. Molyneux, 1985; Moser and Levy, 1986). Meeting practical gender needs is important in supporting the everyday life situations of everybody: women and men as well as individual groups with differentiated needs. Indeed, various feminist researchers in planning sciences assert that the everyday life approach (Lefebvre, 1974/1991) is a useful bridging concept for evaluation of the quality of built environment and open spaces, considering gender issues in spatial sciences (Horelli, Booth and Gilroy, 2000; Jarvis *et al.*, 2009). Especially in the case of Vienna and also in urban planning debates (Gehl, 2010; Gehl and Svarre 2013), many concepts and terms which were first discussed in the context of women-centred and gender-sensitive planning have become mainstream concepts for high-quality planning. In the context of neighbourhood planning, this is the argument for the "city of short distances" or "high quality density", which are compatible terms in urban planning "male-stream" debates. The "target group orientation" or the "fair-shared cities concept" (Fainstein, 2010) in planning are also recognised in spatial development debates. The term "fair-shared city" was used for the branding of the Viennese gender planning approach in 2004.

Guidelines are an important driver to support quality standards

Criteria lists and manuals are the concretisation of a certain planning approach on a general level; this becomes especially clear when qualitative aspects are captured in numbers. However, they remain on a general abstract level, formulating a kind of ideal or standard. The value of their generalisation is comparability and therefore transferability. They give orientation. Mostly they are used to formulate calls for tenders, as a basis of argumentation or to operationalise analysis. In the case of WBS with its formalisation and one single number encapsulating achieved quality, this can also function as a "seal of quality". Manuals or guidelines are a valuable precondition: they help to transfer considerations of general validity and to define the framework and general aims for a specific planning process, they support systematic argumentation when deciding between different proposals, and they define prequalification criteria for competitions.

But manuals are always a kind of short cut. Manuals mirror the state of the art for average "mainstream" situations and solid quality. They are less suitable for innovative specific solutions. In every planning process one has to handle a bundle of often conflicting targets. The decisions are always the outcome of complex considerations that make some aspects more important than others. This results in planning approaches and attitudes that are very often not explicit. So manuals and criteria lists create transparency and provide orientation, but by their very nature can only support the complex and holistic process of planning for a specific situation with all its framework conditions. For several reasons manuals cannot do the "whole job": they are only – although important – working tools that can support mainstreaming processes effectively. However, meeting strategic gender needs challenges the asymmetric power relationship between genders in the sense of gender equality. Indeed, strategic gender-sensitive approaches also mean externalising planning philosophies which support a democratic understanding in planning practice (Moser and Levy, 1986).

Participation of gender experts supports the integration and discourse of gender sensitivity in urban and spatial development processes

When gender experts are involved in calls for tender, make a specific gender prequalification and are members of a jury, they have only one vote among many in the formal decision-making body. The decision of the jury depends to a high degree – especially in the case of different opinions – on the weight gender arguments have for other members. Most importantly: the outcome of an open competition always depends on the planning approaches of the participants.

With open-minded mainstream offices, it is possible to "translate the intention of the tender into a gender-balanced functional programme and good design" in the entry. In the case of the two park competitions, this worked well, also from a gender+ perspective. In some other competitions where gender experts were involved, this structural "weakness" led to less satisfying results, independent of the scale of development or topics. Even in supposedly "gendered" competitions, gender-sensitive criteria are at risk of being of little significance.

As far as the authors know, from a gender planning point of view there has to date been no instance of an ideal-typical process where all those involved in the design and decision-making process had gender+ competence.[14] In the projects realised throughout Europe the impact of the gender planning approach varies with the level of political support for gender issues, the gender planning expertise (and technical competence) of the experts involved, general favourable or unfavourable framework conditions, and the quality of participation processes with regard to their gender sensitivity or blindness. Mainstream competitions with "added gender aspects" do have limitations. In Vienna, however, the participation of a gender expert over many years and gender-quality assessment activities have led to an increased emphasis on open spaces, entrance spaces, staircases and communal spaces in the field of subsidised housing and have improved the quality of parks and playgrounds.

Strategic gender interests still have to be pursued at different political levels and scales of development, also to ensure that manuals or – more importantly – the underlying values and principles of gender-sensitive planning are considered even more intensely in planning practice and research. "Lares" in Switzerland, FOPA in Germany, Urbanistas in the UK, Col·lectiu Punt 6 in Spain – the list is not complete – and other feminist networks contribute a great deal to this work. The steady involvement of gender experts in mainstream planning activities will be necessary so that gender+ aspects will be implemented in the everyday practice of planning. The cases of Vienna, Munich, Berlin or the Ruhr region show (Bauer and Frölich v. Bodelschwingh, 2017) that the establishment of gender-planning expertise within the city administration supports or is favourable to long-lasting mainstream effects, sometimes at the price of becoming less explicit.

Notes

1 In Austria, planning is legally regulated on the federal and municipal levels (Vienna as a city being a federal state in its own right); in Switzerland planning laws and regulations are found on the three levels of state (national), cantons (federal) and municipalities (communal).

2 Figures verified by Wiener Wohnbauforschung (Vienna Research on Housing Construction).

3 Floor area ratio (FAR) is the ratio of a building's total floor area (gross floor area) to the size of the plot upon which it is built.

4 In 1992, a conference at ETH Zurich brought together gender-sensitive perspectives in German-speaking Switzerland. The building department of the Canton of Bern (1992) and the Federal Office for Spatial Planning (1994) publish on gender issues.

Since 1990 ETH Wohnforum, today CASE (Centre for Research on Architecture, Society & the Built Environment) at the Building Department focuses its research on (amongst other topics) user needs esp. in the context of cooperative building associations (first funded project "Housing Situation of Women in the Swiss German Part", 1996). For more information on gender issues in Switzerland see Chapter 2.

5 The objectives set up by "Lares" are supported by objectives for social sustainability as formulated, e.g. in the guidelines of the Swiss associations of engineers and architects SIA "Standard guide to SIA norm 112 – Sustainable construction – Structural engineering" (SIA, 2005) and "Leaflet 2050 and documentation 0246. Sustainable spatial development – Local and regional planning" (SIA, 2015).

6 Infrastructure includes leisure time and recreation facilities (Z 59–73), consumer provision (Z 74–77) and social infrastructure (Z 78–90).

7 For an overview of various planning levels, see Chapter 4 "Gender Mainstreaming as a comprehensive planning strategy".

8 Wohnprojekt as a time-consuming activity and with relatively high costs is a middle-class project. Some aspects of the cohousing project are very interesting from a gender perspective, such as the shared kitchen and dining room on the ground floor, with the children's playroom separated only by a glass wall. Here every weekday an informal organised lunch takes place, consumers can decide to take part from day to day, but have to cook periodically. Some of the inhabitants organise a food coop.

9 The criteria cover the following topics: community, design, usage and local public infrastructure, health and well-being. Questions of mobility are listed in the environment section under the heading of infrastructure. Especially in the field of community, issues of diversity play an important role and can be seen as relevant to gender issues: integration and cultural, social and intergenerational mix, social contact, solidarity and justice, participation.

10 Kalkbreite received the following ratings: residential location 23 of 24, residential complex 31 of 32 and average home of all homes 35.5 of 44. Overall, the development achieved 89.5 of 100 points.

11 Data were collected and exploited according to qualitative research methods, mainly based on personal interviews with inhabitants of different household types, age, sex, living conditions and social/cultural background.

12 There are cities in Switzerland that have gender-sensitive guidelines, e.g. the City of Bern: www.bern.ch/themen/planen-und-bauen/bern-baut/wie-wir-planen-bauen/handbuch.

13

> What has been created is an attractive, functionally multi-faceted district park. The materials used tell of the previous history of the area. The park that was created enables a multitude of uses and acts as a harmonious green space in the area of tension between railway tracks, new housing construction and much-frequented areas in the environment such as the Toni Areal. Play areas have been created that form smooth transitions, plantations that create places of shelter and yet retain clarity, quiet places to sit and access routes with various facilities.
>
> (Hoppe, 2017)

14 Although the gender-sensitive competition organised by the Coordination Office Vienna for the redesign of two small neighbourhood parks came close to it – as the gender experts invited for a preparation workshop wrote the call of tender, made the prequalification and proposed members of the jury.

References

Bauer, U. and Frölich v. Bodelschwingh, F. (2017). *30 Jahre Gender in der Stadt- und Regionalentwicklung (30 years of gender in city and regional development)*. Berlin:

Deutsches Institut für Urbanistik (Difu). Available at: https://difu.de/publikationen/2017/30-jahre-gender-in-der-stadt-und-regionalentwicklung.html [Accessed 9 Aug. 2018].

BWO – Bundesamt für Wohnungswesen (Ed.) (2015). *Wohnbauten planen, beurteilen und vergleichen. Wohnungs-Bewertungs-System WBS. Ausgabe 2015 (Planning, evaluating and comparing residential buildings. Housing evaluation system. Edition 2015).* [Online] Grenchen. Available at: www.wbs.admin.ch/de/www.wbs.admin.ch/fr www.wbs.admin.ch/it [Accessed 9 Aug. 2018].

BWO – Bundesamt für Wohnungswesen (Ed.) (2000). *Wohnbauten planen, beurteilen und vergleichen. Wohnungs-Bewertungs-System WBS. Ausgabe 2000 (Planning, evaluating and comparing residential buildings. Housing evaluation system. Edition 2000).* Schriftenreihe Wohnungswesen, Bd. 69. [Online] Grenchen. Available at: www.wbs.admin.ch/de/www.wbs.admin.ch/fr www.wbs.admin.ch/it [Accessed 9 Aug. 2018].

BWO – Bundesamt für Wohnungswesen (Ed.) (1986). *Wohnungs-Bewertung. Wohnungs-Bewertungs-System WBS. Ausgabe 1986 (Housing evaluation system. Edition 1986).* Schriftenreihe Wohnungswesen, Bd. 35. [Online] Bern. Available at: www.wbs.admin.ch/de/www.wbs.admin.ch/fr www.wbs.admin.ch/it [Accessed 9 Aug. 2018].

BWO – Bundesamt für Wohnungswesen (Ed.) (1979). *Wohnungs-Bewertung in der Anwendung (Applied housing evaluation).* Schriftenreihe Wohnungswesen, Bd. 9. Bern.

Damyanovic, D., Reinwald, F. and Weikmann, A. (2013). *Gender mainstreaming in urban planning and urban development.* Werkstattbericht, Nr. 130A. [Online] Vienna: Urban Development Vienna, Municipal Department 18 (MA 18) – Urban Development and Planning. Available at: www.wien.gv.at/stadtentwicklung/studien/pdf/b008358.pdf [Accessed 9 May 2018].

Deutsche Gesellschaft für Nachhaltiges Bauen (DGNB) (undated). [Online]. Available at: www.dgnb.de/de/ [Accessed 9 Aug. 2018].

durable Planung und Beratung GmbH and raumdaten GmbH (2013). *Monitoring Nachhaltigkeit Siedlung Kalkbreite. Gesamtdokument Planung und Bezug (Monitoring sustainability development Kalkbreite. Overall documentation planning and move-in phase).*

Eidgenössische Forschungskommission Wohnungsbau. Der Delegierte für Wohnungsbau (Ed.) (1975). *Wohnungs-Bewertungs-System (WBS) (Housing evaluation system).* Schriftenreihe Wohnungsbau 28d. [Online] Bern. Available at: www.wbs.admin.ch/de/www.wbs.admin.ch/fr www.wbs.admin.ch/it [Accessed 9 Aug. 2018].

Emmenegger, B., Fanghänel, I. and Müller, M. (2017). *Nachbarschaften in genossenschaftlichen Wohnsiedlungen als Zusammenspiel von gelebtem Alltag, genossenschaftlichen Strukturen und gebautem Umfeld – Ein Beitrag zur sozialen Nachhaltigkeit (Neighbourhoods in cooperative housing developments as interaction between everyday living, cooperative structures and built environment – a contribution to social sustainability).* Schlussbericht zum KTI-Forschungsprojekt 2014–2017 der Hochschule Luzern – Soziale Arbeit. [Online]. Available at: www.hslu.ch/de-ch/hochschule-luzern/forschung/projekte/detail/?pid=1005 [Accessed 9 Aug. 2018].

Fainstein, S. (2010). *The just city.* Ithaca, NY: Cornell University Press.

Gehl, J. (2010). *Cities for people.* Washington, Covelo, London: Island Press.

Gehl, J. and Svarre, B. (2013). *How to study public life. Methods in urban design.* Washington, Covelo, London: Island Press.

Genossenschaft Kalkbreite (undated). [Online]. Available at: www.kalkbreite.net/ [Accessed 9 Aug. 2018].

Grün Stadt Zürich (2010). *Pfingstweid, Quartierpark mit Schulanlage Zürich-West, Projekt- und Ideenwettbewerb. Bericht des Preisgerichts (Pfingstweid, district park with school complex, Zurich-West, project and concept competition – Report of the Panel of Judges).*

Hager, G., Altermatt, P. and Posset, P. (Eds.) (2008). *Rudolf-Bednar-Park Wien (Rudolf-Bednar-Park Vienna)*. Salenstein: niggli Verlag.

Hoppe, G. (2017). *Factsheet Pfingstweidpark*. Lares.

Horelli, L., Booth, C. and Gilroy, R. (2000). *The EuroFEM toolkit for mobilizing women into local and regional development*. [Online] Helsinki: Helsinki University of Technology, Centre for Urban and Regional Studies. Available at: https://issuu.com/eva_alvarez/docs/12_eurofem_toolbox [Accessed 4 Aug. 2018].

Irschik, E. and Kail, E. (2013). Vienna. Progress towards a fair shared city. In: I. Sánchez de Madariaga and M. Roberts (Eds.), *Fair shared cities. The impact of gender planning in Europe*. Farnham: Ashgate, 193–230.

Jarvis, H., Kantor, P. and Cloke, J. (2009). *Cities and gender*. London: Routledge.

Lares (Ed.) (2012). *Mit Lares bauen und planen. Was es Ihnen bringt (Building and planning with Lares. What are your benefits?)*. Bern: Lares.

Lares (2008). *Bedarfsabklärung Gender für das Wohnungs-Bewertungs-System WBS (Checking the housing evaluation system for gender perspectives)*. Bern: Lares.

Lefebvre, H. (1974/1991). *The production of space*. Translation of: *La production de l'espace*. Malden, MA, Oxford and Victoria: Blackwell Publishing.

Mehr als Wohnen (2007). *Gemeinnütziger Wohnungsbau in Zürich. 1907–2007 (Not-for-profit housing construction in Zurich. 1907–2007)*. Zurich: gta.

Molyneux, M. (1985). Mobilising without emancipation. *Feminist Studies*, 11(2), 227–254.

Moser, C. and Levy, C. (1986). *A theory and methodology of gender planning. Meeting women's practical and strategic needs*. DPU Gender and Planning Working Paper No. 11, Development Planning Unit, Bartlett School of Architecture and Planning University College London.

Netzwerk Nachhaltiges Bauen Schweiz NNBS (2016). *Standard Nachhaltiges Bauen Schweiz SNBS. Version 2.0 (Swiss standard for sustainable building. Version 2.0)*. [Online]. Available at: www.nnbs.ch/fileadmin/user_upload/SNBS_Download/SNBS_2-0-Kriterienbeschrieb-Hochbau-d-180112.pdf [Accessed 9 Aug. 2018].

SIA – Schweizerischer Ingenieur- und Architektenverein (2015). *SIA-Merkblatt 2050 und Dokumentation 0246. Nachhaltige Raumentwicklung – Kommunale und regionale Planung (SIA – Swiss association of engineers and architects. Leaflet 2050 and documentation 0246. Sustainable spatial development – Local and regional planning)*. Zürich: SIA (www.sia.ch).

SIA – Schweizerischer Ingenieur- und Architektenverein (2009). *Norm SIA 500. Hindernisfreie Bauten. Korrigenda SIA 500/C1 (2009), SIA 500/C2 (2011), SIA 500/C3 (2013) (SIA – Swiss association of engineers and architects. Norm 500. Buildings without barriers)*. Zürich: SIA (www.sia.ch).

SIA – Schweizerischer Ingenieur- und Architektenverein (2005). *SIA 112/1. Nachhaltiges Bauen – Hochbau – Verständigungsnorm zu SIA 112. (SIA – Swiss association of engineers and architects. Standard guide to SIA norm 112 – Sustainable construction – Structural engineering)*. Zürich: SIA (www.sia.ch).

Siegel, R. (2012). *Nutzenevaluation Lares. Schlussbericht zuhanden des Projektes Lares (Evaluation Lares. Final report for the Lares project)*. Bern: Zentrum für Bildungsevaluation, Pädagogische Hochschule Bern.

Stadt Zürich (undated a). *Pfingstweidpark*. [Online]. Available at: www.stadt-zuerich.ch/ted/de/index/gsz/natur-_und_erlebnisraeume/park-_und_gruenanlagen/pfingstweidpark.html [Accessed 10 Oct. 2017].

Stadt Zürich (undated b). *Portal. Wohnpolitik (Platform housing policy)*. [Online]. Available at: www.stadt-zuerich.ch/portal/de/index/politik_u_recht/stadtrat/weitere-politikfelder/wohnpolitik.html [Accessed 9 Aug. 2018].

Stadt Zürich – Liegenschaftenverwaltung (2016). *Geschäftsbericht (Annual report)*. [Online]. Available at: www.stadt-zuerich.ch/fd/de/index/liegenschaftenverwaltung/Publikationen/geschaeftsbericht_2017.html [Accessed 9 Aug. 2018].

Stadtentwicklung Wien – Magistratsabteilung 18 (MA 18), Stadtentwicklung und Stadtplanung (Urban Development Vienna – Municipal Department 18, Urban Development and Planning) (2018). *Fachkonzept Öffentllicher Raum (Thematic concept. Public space)*. Werkstattbericht, Nr. 175. [Online]. Available at: www.wien.gv.at/stadtentwicklung/studien/pdf/b008523.pdf [Accessed 9 May 2018].

Stadtentwicklung Wien – Magistratsabteilung 18 (MA 18), Stadtentwicklung und Stadtplanung (Urban Development Vienna – Municipal Department 18, Urban Development and Planning) (2017). *Städtebauliche Kennwerte – Analyse in Praxis und Modell mit Schwerpunkt Freiraumplanung (Manual of Urban Development Indicators – Analysis of practice and models on the topic of open space planning)*. Werkstattbericht, Nr. 167. Vienna.

Stadtentwicklung Wien – Magistratsabteilung 18 (MA 18), Stadtentwicklung und Stadtplanung (Urban Development Vienna – Municipal Department 18, Urban Development and Planning) (2013). *Evaluierung Rudolf-Bednar-Park – Nutzungsevaluierung (Rudolf-Bednar-Park – Evaluation of utilization*. Werkstattbericht, Nr. 138. [Online]. Available at: www.wien.gv.at/stadtentwicklung/studien/pdf/b008348.pdf [Accessed 31 Oct. 2018].

Swiss Federal Statistical Office (undated a). *Rented dwellings*. [Online]. Available at: www.bfs.admin.ch/bfs/en/home/statistics/construction-housing/dwellings/rented-dwellings.html [Accessed 9 Aug. 2018].

Swiss Federal Statistical Office (undated b). *Wohnverhältnisse nach Bewohnertyp für die 10 grössten Städte kumuliert 2014–2016 (Cumulated dwelling standards for inhabitant types for the 10 biggest cities 2014–2016)*. [Online]. Available at: www.bfs.admin.ch/bfs/de/home/statistiken/bau-wohnungswesen.assetdetail.4742962.html [Accessed 9 Aug. 2018].

Vienna City Administration – Municipal Department 18 (MA 18), Urban Development and Planning (2015a). *STEP 2025. Thematic concept. Urban mobility plan Vienna. Together on the move*. [Online]. Available at: www.wien.gv.at/stadtentwicklung/studien/pdf/b008443.pdf [Accessed 31 Oct. 2018].

Vienna City Administration – Municipal Department 18 (MA 18), Urban Development and Planning (2015b). *STEP 2025. Thematic concept. Green and open space. Sharing the outdoors*. [Online]. Available at: www.wien.gv.at/stadtentwicklung/studien/pdf/b008440.pdf [Accessed 31 Oct. 2018].

Vienna City Administration – Municipal Department 18 (MA 18), Urban Development and Planning (2014). *STEP 2025. Urban development plan Vienna.* [Online]. Available at: www.wien.gv.at/stadtentwicklung/studien/pdf/b008379b.pdf [Accessed 9 May 2018].

Vienna City Administration – Municipal Department 23 (MA23), Economic Affairs, Labour and Statistics (2017). *Vienna in figures*. [Online]. Available at: www.wien.gv.at/statistik/pdf/viennainfigures-2017.pdf [Accessed 9 Aug. 2018].

wohnfonds_wien – Fonds für Wohnbau und Stadterneuerung (undated). [Online]. Available at: www.wohnfonds.wien.at/ [Accessed 9 Aug. 2018].

6 Evaluation of spatial development from the gender+ perspective

A methodological proposal

Liisa Horelli and Doris Damyanovic

In spite of the many historical waves of enhancing equal opportunities, there are relatively few examples of engendering urban planning and community development. Consequently, there is ambivalence about the significance of gender in spatial planning, particularly as gender mainstreaming (GM[1]), without positive action, seems to increase the importance of diversity issues and diminish those of gender equality.[2] However, mainstream evaluation theories and practices are mostly gender blind, unless a gender perspective is explicitly mandated when the evaluation is commissioned (Bustelo, 2017). This is a rare occurrence especially in urban planning. Admittedly, conducting evaluations of spatial development from a gender perspective is challenging, since gender concepts applicable to such contexts are ambiguous, urban planning policies and systems vary considerably from one country to another, and the selection of an appropriate evaluation approach among the bewildering diversity of options on offer is far from self-evident.

In addition, there are many types of urban planning contexts which require different kinds of engendering and evaluation in order to have desired results. The best known examples of gender mainstreaming in urban planning and planning evaluations come from Austria, especially from Vienna, but other cases of different types of engendering also exist (Horelli, 2017; see Chapter 1).

Consequently, it is difficult to provide evidence of the importance of gender in spatial planning and to demonstrate its significance. Therefore, the research question for this chapter concerns how to monitor and evaluate the process of spatial development and the outcomes from the gender+ perspective at the project level.

The aim of this chapter is to present and discuss the methodological results of gendered evaluations of two cases in the context of self-organisation around urban planning and spatial development, since the other examples in this book represent top-down gender mainstreaming. Both cases come from Helsinki (650,000 residents), the capital of Finland (6.5 million population), where there are no examples of gender mainstreaming but numerous self-organising groups directly or indirectly associated with spatial development. In fact, conspicuous social change seems to be taking place in Helsinki, as the adoption of information and communication technologies (ICTs) and especially social media enables citizens to act directly upon environmental issues, to organise different activities and events, and to co-produce peer-to-peer services and a variety of goods, which have spatial impacts. The latter activity contributes to the sharing or solidarity economy with alternative currencies, which is endorsed by coalitions or networks, such as Facebook (FB)-enhanced flea-market groups, as well as local food circles and co-ops, which apply crowdfunding for their new businesses.

The diversity of the content and forms of urban activism or commoning[3] in Helsinki has been categorised by Mäenpää and Faehnle (2017a) as citizen economy services (food groups, time banks, neighbourhood help), community activism (events, local and social movements, social peer support), digital activism (hackathons, development of digital society), activism support (innovation and transmission communities, tools) and space-related activism (urban planning groups, squatters, block yards, living room exhibitions, murals and graffiti). FB groups often have a spatial impact as they provide virtual space and tools to shape the physical space. This has gradually led to a dialogue with the city administration that can be called a form of co- or hybrid governance (Horelli, 2013; Puerari, 2015; Mäenpää and Faehnle, 2017b).

Surprisingly, most urban activist groups are women-led or some 70 per cent of their members are female, such as the organisers of the "Pop-up Restaurant day" or the "Cleaning day", when the whole city is full of ad hoc restaurants or the city is turned into a flea-market (Horelli *et al.*, 2015). Nevertheless, this does not mean that the movements are gender sensitive or feminist. Finland is a gender-neutral[4] or even a gender-blind country in many fields, although Finnish women were the first in the world, in 1906, to be granted both the right to vote and the right to stand in elections (see Chapter 2).

The chosen evaluation cases represent space-related activism (Mäenpää and Faehnle, 2017a). The first one is called "More City to Helsinki" (MCH), which is a male-led FB group that seeks to densify and enlarge the core city area in the image of Berlin. The second one is the "National Urban Park to Helsinki-movement" (NUPH), a women-dominated group that strives to protect not only the Helsinki Central Park but also important cultural, eco-social and marine settings.

We argue that the complexity of engendering evaluations of urban planning requires an integrative evaluation framework and the adoption of mixed methods in order to facilitate knowledge building and activate "learning to learn" behaviours in the management of urban issues. The aim is therefore to base the evaluation methodology on a transdisciplinary theoretical framework, adopting a carefully chosen value position (equity, agency, voice) in the design of the endeavour, which applies participatory methods leading eventually to evaluative thinking about the meaning of gender at large.

The theoretical framework of the chapter is presented in the next section, followed by the evaluation cases and their comparison. We conclude by discussing the methodological question concerning ways to carry out gendered evaluations of spatial development.

Constructing an integrative framework

As evaluations of urban planning or spatial development from a gender perspective are not common, there is no established methodology nor best practice. Consequently, there is a need to construct an integrative framework comprising, for example, concepts from planning theories, feminist/gender studies and evaluation approaches. As the introduction to this book in Chapter 1 already describes relevant gender concepts, we will focus, besides positioning urban planning as a gender-blind context, on engendering the evaluation methodology.

Urban planning as a gender-blind context

According to Nadin and Stead (2008), two major types of spatial planning systems exist in Europe. On the one hand, the continental, imperative type is dominant in the Nordic, Germanic and Roman countries. It is based on a set of abstract rules and principles that are applied at the outset of urban planning. On the other hand, the Anglo-Saxon system is permissive and evolves based on precedent cases. These two types have different foci and systems of coordination as well as consequences. To date, both systems are gender blind. However, European Union practices seek to make the systems converge within a broad framework of planning principles (Reimer, Getimis and Blotevogel, 2014), some of which have been impacted by gender mainstreaming, as described in several chapters in this book.

Due to the different planning systems and theories, a variety of urban planning definitions coexist, ranging from pragmatic arrangements of available physical space to the "organising of hope" in the design of human settlements (Hillier and Healey, 2008). The complexity of the latter increases the challenges of evaluation. However, most evaluations in urban planning are in practice conducted *ex ante*, before the plans have been implemented, just like in environmental or social impact assessments, gender impact assessments (GIA) included. There are very few holistic evaluations of urban plans and their implementation, or of their *ex-post evaluation* of outcomes with learning and accountability purposes, particularly from the gender perspective (Bustelo, 2017).

A feminist or gender+ perspective in the evaluation of urban planning?

Gender is not just about sexual identity. It is a psycho-corporal and socio-cultural construction of masculinities and femininities embedded in a system of power relations, reflected in a certain gender order or in the gender contract (see Table 1.1 in Chapter 1). The latter has to be deconstructed to reveal its instrumental significance in different contexts. This means that engendering can be applied from different perspectives and with various models, such as the liberal, gender-neutral perspective (inclusion), the radical women's perspective (reversal), or the postmodern gender perspective in which gender categories (women and men) are deconstructed and the focus is on the transformation of gender structures (Squires, 2005). However, according to Arturo Escobar (2017), the dominating world view – patriarchy, which is embedded in capitalist markets – represses both women and men to the extent that it is only possible to achieve limited equality within and between genders. He leans towards another approach to life – radical interconnectedness – which is based on an ontological conception called "the biology of love". Instead of valuing competition, hierarchies and growth, this framework underlines care of humans, non-humans and nature, collaboration, inclusion and participation, which form the basis for the emerging feminist politics of solidarity (Libreria delle donne di Milano, 1989, 1996). This can be seen, according to Julie Matthaei (2018), as a shift from identity politics and the inequality paradigm to that of solidarity.

Whatever the perspective is, the interrelation between space, gender and varying types of power cannot be separated, as they are intertwined in the policy and planning processes (Damyanovic, 2013; Wotha, 2013). Consequently, gendered evaluations have to focus on the roles, responsibilities, resources and interests/visions of men and women from different backgrounds in a given context (see also Mies, 1994).

Although all evaluation approaches have their own implicit or explicit values, feminist evaluations are particularly transparent concerning their core values: equity/equality, voice and agency. However, feminism has been defined as a cluster of contesting views on the general problematic (Verloo, 2007). Donna Podems (2012) claims that unlike most gender approaches, feminist evaluation does not provide a precise framework but rather a way of thinking about evaluation. It implies that the focus is on the gender inequities leading to social injustice; gender inequities are seen as systemic and structural; evaluation is regarded as a political activity; and knowledge as well as different ways of knowing are a resource for the people sharing it (Sielbeck-Bowen *et al.*, 2002).

Nevertheless, evaluations from the gender+ perspective currently share several feminist features. This means that they also challenge the subordinate position of women and accept that women are different from men, even if they are not a homogeneous group but are influenced by intersectionality, including varying sexual orientations. Why are women and men different or why do they have divergent interests in urban planning, might also be a question of concern. However, as not all gender-sensitive evaluations actively advance the feminist values of equity, agency and voice, feminist and gender-sensitive evaluations can be regarded, although not as the same, as complementary. Nevertheless, this implies, like Maria Bustelo (2017: 97) claims, that "feminist evaluation is a necessary watch dog that prevents the watering down of the evaluation from the gender perspective".

Designing a gendered evaluation with iterative steps

Evaluation designs generally depend on the context and purpose, and the gendered ones are no exception. Gendered evaluations of spatial development, urban planning including, require a design characterised by several iterative steps. These comprise the definition of the purpose and object of the evaluation, its resources, contextual analysis, application of programme theories (change and action theories), the choice of evaluation questions and criteria that can then be measured by gender-relevant indicators, the selection of methods and finally the dissemination of results (Horelli and Wallin, 2010). Reflection on ethical principles, such as independence, transparency, impartiality, quality and intentionality, is also part of the ongoing process (BetterEvaluation, 2017).

Ideally, engendering evaluation of (urban) planning and development refers to the systematic and systemic determination of worth or merit from a gender+ perspective. This involves an examination of how gender intersects with other inequalities, such as age, class, ethnicity, disabilities and sexual orientations.

Theory-driven evaluations (for example Chen, 2005) are important in the assessment of (spatial) planning and development, because they enhance the comprehension of the core questions, such as what is sought after (visions and expected results), and why or with which mechanisms the interventions affect the results, as well as how to achieve the outcomes. Besides being theory-driven, gendered evaluations frequently embody different evaluation approaches, such as utilisation and equity-focused empowerment evaluation or evaluation 2.0 that applies tools from social media, to mention just a few.

A qualitative meta-analysis of several urban projects at the GDUS-seminar (European Network on Gender and Diversity in Urban Sustainability), in Hannover in April 2012, indicated that there is a need for gender-sensitive criteria for both the participatory process of (urban) planning and its content (Horelli, 2009, 2012). The choice of process criteria is relatively easy, as this concerns the organisation and the selection of modes and tools of participation, as well as its impact on various levels. The choice of relevant content criteria, however, is quite difficult, due to the variety of planning contexts and purposes. One possibility is to identify a few core planning issues that interact with a set of cross-cutting criteria producing indicators for (urban) planning content, which in turn measure the criteria (see Table 6.1 and Chapters 4 and 5).

Table 6.1 is based on the following principles: (1) Strengthening the city of short distances; (2) Targeting the daily requirements of different groups, especially those that are dependent on their local environment.[5] As the environment comprises both physical and socio-cultural elements (Fleury-Bahi, Pol and Navarro, 2017), the criteria in Table 6.1 vary in terms of the extent to which the element is interpreted as physical or socio-cultural (those on the left-hand side are more physical, whereas the socio-cultural aspects increase to the right).

The choice of mixed methods

Evaluation from the gender+ perspective tends to apply a set of mixed qualitative and quantitative methods, which was also the case in the examples in this chapter.

Table 6.1 Matrix of content criteria and indicators for a comprehensive (urban) planning and spatial development strategy on different planning scales (e.g. concepts and strategies for design, master plan, land-use plan)

Cross-cutting criteria/ Planning issues as the interplay of:	*Diversity and flexibility*	*Accessibility*	*Safety and security*	*Image, scale and representation*	*Co-governance*
Production and reproduction	Balanced ratio of living, working and socialising; a network of traditional, individual and collective work/living spaces; supply of goods and services, recreational and cultural activities	Easy access to services and work due to proximity, connections, opening hours and ICTs; barrier-free access to different types of spaces	Recognition of both physical and perceived safety (the role of social control and physical barriers)	Spaces/places for users of different genders, ages, class, ethnic background, sexual orientation; diversity of cultural, artistic and work settings	Co-governance, including time-planning and co-production via simultaneous examination of work, housing, services, mobility and recreation
Private–public spheres	Allocation and design of different spatial scales and the nature of private–public and transitional, in-between open spaces	Clear socio-cultural rules regarding hierarchy and nature that can be negotiated; time policies and planning; glocal access through ICTs	Vandalism is immediately dealt with; social control according to the nature of the spheres	Well-kept and personalised communal and "political" spaces; symbolic signs of multiple genders; sense of community	Self-organising around different spaces for building community; co-governance of the local-global connections

Mobility and communication systems	A network of linked (linear and radial) transportation modes organised by PPPPs; a variety of mobility services enhanced by ICTs and flex street network	Pedestrian and bicycle pathways to neighbourhood services; public transportation linked to the external mobility networks	Well-lit and safe stations, bus stops at night; well-designed parking lots; cyber-security of ICTs	Cool and comfortable public transport, bicycle and pedestrian infrastructure with up-to-date ICTs	Citizen panel of women and men to monitor the mobility system and mobile services
Housing solutions	A mix of different types of dwellings (individual, row, high-rise; private, rental) and formats (cohousing, special housing)	Barrier-free planning; long-term living by access to additional rooms, storage, or work spaces	Healthy building materials; resident control and collaboration	Long-term dwelling and communal spaces reflected in the personalisation of the buildings and yards; attractive with human scale	Self-organisation of residents who have a say in the maintenance and co-governance of the neighbourhood
A variety of natural environments	A mix from nature-based to more designed landscapes and soundscape settings (with silence/vitality)	Barrier-free proximity to different kinds of green spaces with varying activities	Lighting and clear views; social control	Beautiful landscapes for recreation and innovative uses (guerrilla gardening, etc.)	Commoning the environment; ecological building and living solutions; gendered space making

The data gathering methods in this chapter comprised individual F2F and telephone interviews, focus groups, an internet survey (Webropol), observation of events and meetings, and analyses of the internet data (FB group, websites), documents and reports. The methods of data analysis consisted of an integration of the 4R-method[6] (JämStöd, 2007) and Critical Frame Analysis[7] (Goffman, 1974; Verloo, 2007) that assisted in the application of content analysis, based on grounded theory and the constant comparative method (Strauss and Corbin, 1990). The categories emerged not only from the critical questions of these two methods but also from the empirical characteristics of the cases (see Table 6.2). Interpretation of the data in the content analysis was enhanced by the application of the integrative theoretical framework as part of the grounded theory. The interpretation of the results and the drawing of conclusions were supported by the so-called Quasi-Judicial (QJ) case method, developed by Bromley (1986). This is based on the network of empirical facts, relations and relevant concepts that was especially useful for the discussion.

Evaluation of two cases in Helsinki from the gender+ perspective

The purpose of the two evaluations was to serve as a methodological proposition, illustrating how an evaluation of spatial development from the gender+ perspective can be carried out at the project level. The nature of the evaluation was pragmatic with some transformative elements. The control of the evaluation process was in the hands of the authors but the design was participatory (Figure 6.1). A variety of participation modes and tools were

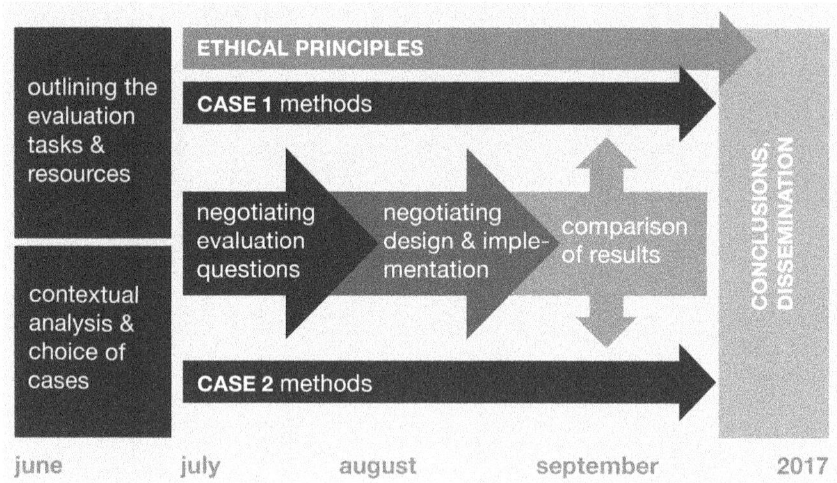

Figure 6.1 The evaluation design of the two cases of self-organisation in Helsinki.
Source: © Horelli.

applied. In practice, the object of the evaluations, carried out during a four-month period in the summer 2017, was the process and outcomes, including their history in retrospective, of the self-organising movements around spatial development and the ways these were related to gender+. Although the evaluations were not commissioned by the movements, the design, main questions and outcomes were negotiated and discussed with core members, and disseminated to the network. The resources for the evaluation comprised the work of the authors, supported by a Webropol assistant.

The design of the evaluations (Figure 6.1) followed the iterative steps described in the previous section. Also, the chosen methodological package was the same as in the preceding part. The exception was that the analysis of FB discussions was only carried out with Case 1, because Case 2 did not have an active FB group, and therefore a Webropol survey was applied.

The evaluation questions were: How was the solution planned and implemented? Who has voice, agency and power? How is gender+ done in the process and reflected in the outcomes? The criteria for the process were derived from the questions, including effectiveness (the achievement of goals). The content criteria of the outcomes (see Table 6.1) centred mostly on "image and scale" as well as "diversity" in the context of the interplay of production and reproduction, as well as in those of the mobility systems and natural environments.

The cases are first described and then the results compared for learning purposes.

Case "More City to Helsinki" (MCH)

The self-organising movement "More City to Helsinki" (MCH) was created in 2009 by its founding father Mikko Särelä (PhD, now 35), who created a political discussion group on Facebook. It took several years to blossom, but by 2013 it had 1,500, in 2014 6,000, in 2015 10,000 and in 2017 over 15,000 members. Until two years ago, the majority of the members were white, well-educated middle-class males (70/30 according to Laaksonen and Porttikivi, 2016), but currently the membership of women and men is around 50/50. However, MCH can be seen as a male-dominated movement, since the seven moderators and core members are all male and only 10–15 per cent of the postings are made by women who are also white, well-educated and from the middle class. The members of this influential movement include many talented participants, who function on a voluntary basis. They are architects, planners, ecologists, urban activists and politicians, who are mostly in their 30s and 40s.

The initial motivation for the creation of the group was ongoing discussion about public transportation, services and, above all, the small urban centre of Helsinki, the capital of Finland. At that time the problem was defined, and still is, as the strictly regulated low density of the city centre leading to a lack of affordable housing in this district.

The vision of a good city varied among the male and female members of the movement. Men tended to stress more urban preferences, such as a dense urban centre providing affordable housing, a fluid mobility system based on rails and cycling, and services, optionality and buzz. One respondent dared to defend spaciousness, but a group of angry male posters found that too much space is counterproductive as "a good city is like Berlin".

Female respondents also endorsed services, the mixture of functions, optionality and walkability. Nevertheless, they also stressed safety, aesthetics and nature. Some of them found that a dense city also suits families with children and older people, due to the lack of car dependence. Taking care of the vulnerable was also part of a good city (see the criteria in Table 6.1). The solution to the problem was seen as planning and building more densely[8] in the centre of Helsinki, a small peninsula which, according to the movement, should be doubled in size increasing the population to one million inhabitants. Thus, the theory of change in MCH seems to imply that by increasing the density of the new master plan, building can spread out both horizontally and vertically, which in turn will enable people to have access to affordable housing in the expanded city centre.

MCH does not have an explicit strategy and the implementation of the solutions takes place by applying a wide set of methods. The theory of action can thus be called "tactical urbanism" (Lydon and Garcia, 2015) in which various short-term actions at the local level will enable long-term change. The methods comprise, above all, the moderation of the FB group by seven male administrators. They apply explicit rules that are announced at the front of the FB site. The purpose of the group is not to discuss the goal – to construct more densely built urban blocks – but to reflect on how the goal can be achieved. The rules also advise avoiding too harsh communication, but many women feel that they do not dare to speak as they do not have sufficient expert knowledge and the right terminology. In addition, they feel insecure, as sometimes they are overridden by the male participants' know-how and sarcasm. Another method of implementation is the monthly pub event, which is frequented more by men than women. Here again, it is the communication style that bothers women, as dialogue may begin with something like: "What do *you* think about this issue?".

The third tactic is to let individual or group ideas flow by encouraging activists to directly contact the city officials, for example on some planning issue, and then to elaborate the proposal in the FB group. It can subsequently be further worked out by organising a "planning walk" in the area, where the idea is discussed in situ. Then, for example, the Urban Helsinki collective (a group of FB professional members), can make a proposal to the City, such as the "alternative master plan" Pro Helsinki 2.0, based on tram-mobility, high densities, turning the highways leading to the city into boulevards, and proposing, for example, to build on all waterfronts and to reduce the designated nature areas in the Helsinki peninsula (see Figure 6.2).

The application of social media as well as the press and TV is a natural part of the arsenal. Nevertheless, one of the most effective methods is networking and lobbying with high-ranking officials and politicians – many of whom are members of the movement. The politicians then discuss the issues at stake in

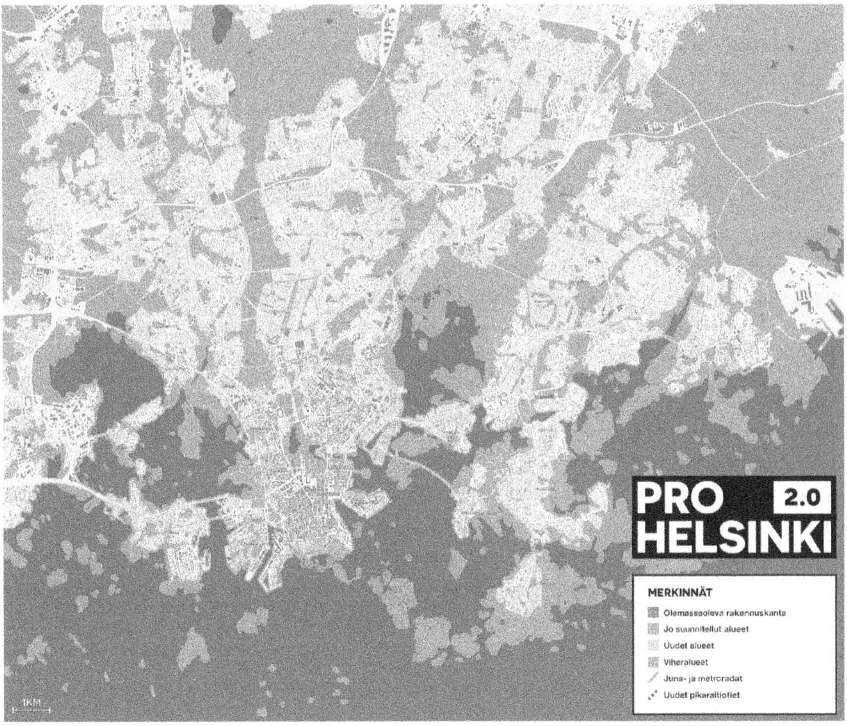

Figure 6.2 "Alternative master plan" Pro Helsinki 2.0. Created by Urban Helsinki (CC-BY-NC) in 2016.

Source: © Urban Helsinki.

their respective parties. The so-called hybrid governance with the City officials informally takes place through all these means. In fact, the City Planning Department has recognised the significance of the group as a resource for planning (Mäenpää and Faehnle, 2017a).

MCH also has international connections from which it draws inspiration for spatial development. Some of its members are engaged in organising the yearly YIMBYcon (Yes in my backyard) event in Helsinki in order to try out new ideas in the local landscape and also to profile the capital as a YIMBY city. The formation of the movement seems low and network-like. However, it has a partial organisation, constructed through the meta-conversation and communicative culture of the administrators, which guides the urban ideology and strategy adopted by the group (Laaksonen and Porttikivi, 2016). For example, criticism and negative comments concerning the master plan, such as commenting on the lack of child-centred and human-friendly values, were forbidden in the group during lobbying for the new master plan. Discussion on immigration and disability is also restricted. The moderators have an internal chat, which enables them to decide who or what can be removed or virtually awarded with a Parrot, the symbol for promoting the goals of the movement (Laakso, 2015). Thus, the flat, partial network-organisation comprises several layers: the administrators (all male), core activists (50–100 mostly male) and the followers.

The outcomes comprise the impacts that MCH has made on the new master plan (2016), which can be seen in the increased density, tram-based mobility and city boulevards, and on the new 2017 Helsinki strategy. The key success, according to the founding members, is the emerging discussion on town planning issues that is no longer in the hands of experts, politicians and officials, but in those of "youngish", mostly male citizens.

Gender is not explicitly salient in the movement, except for specific events, such as "What is feminist urban planning", organised in spring 2016. Women complain from time to time that they do not have a voice in the discussion due to the communication culture of the group. Expert knowledge is applied to undermine experiential knowledge, although the basis of expertise is often questionable, at least in the case of what urbanism, urban culture and space mean.

Agency seems to lie in the hands of the cycling male. The core tasks and most postings are carried out by men (Table 6.2). Despite this conspicuous gender bias, the administrators have not taken action to repair it. This has consequences for reduced opportunities to opt for a discussion on, for example, the integration of density with quality requirements. Besides gender, there is very little discussion or interest in age, disability, ethnicity or immigration-related issues, nor in sexual orientation, all of which might have spatial impacts.

Case "National Urban Park to Helsinki" (NUPH)

The "National Urban Park to Helsinki" (NUPH) movement was created in February 2015 as an F2F group of 50 members with common interests, as the members were tired of the false promises of the City officials. Currently, it is a self-organised movement comprising around 100 active, mostly white, middle-class, well-educated members (two-thirds female), as well as 7,000 citizens with 80 neighbourhood and other organisations, who signed a petition to found a National Park.

The main problem for the members was the drafting of the new, highly effective master plan that threatened to reduce the size of the Central Park and several important ecological and socio-cultural settings. Thus, the new master plan will restrain the eco-social development of the city and destroy many meaningful places in various neighbourhoods.

The vision of a good city for NUPH members is holistic and multidimensional (see the criteria in Table 6.1). It has a human scale (not too high and dense), natural elements (biodiversity in woods, parks, seashore), cultural aspects, services, flexible public transportation and dwelling options. It represents a child-friendly and safe environment.

The suggested solution to the problem was to adopt and implement the concept of the National Urban Park (NUP) as decreed in the 9th chapter of the

Finnish Land Use and Building Act of 2000. NUP is an ensemble of precious urban ecological and socio-cultural settings that narrate the story of the city. There are eight of them in Finland.[9] It is a unique concept, which received the European Garden Award in 2014. Consequently, the theory of action in this case means that the adoption of an NUP in Helsinki might contribute to the development of a good city. At least it will protect the Central Park and some of the fragile cultural settings threatened by the new master plan.

A month after the founding of the movement, the members co-produced their strategy. Its mission was to found an NUP by creating a network of blue, green and cultural environments in Helsinki. The values in the strategy comprise freedom (to support all neighbourhoods) and modesty (not all settings can be chosen in the first phase). The ethical principles are legality (cf. the Land Use and Building Act), independent expertise, one for all and all for one, and the adoption of a positive attitude.

The means for implementing the solution are juridical, political and organisational. Networking with all possible partners – politicians, officials, researchers and students, and local, national and international associations – is at the core of the tactics. The strategy is operationalised by a dynamic plan of action that is updated almost weekly and disseminated on the website, which functions as the main information channel. The FB group is not very active but regular F2F meetings with 20–30 people take place every fortnight in which the tasks, such as meetings with politicians, preparing statements, organising events and participation, for example, in the formal feasibility study of the NUP, are distributed. Thus, the theory of action implies that a positive, soft approach to negotiations with different stakeholders will bring forth the desired results.

There is no formal organisation, but about a dozen founding members from the core group, most of whom are women, undertake the daily or weekly work. Around 20–60 activists show up interchangeably at the events, although there were, for example, 80 people at the kick-off meeting of the feasibility study. The values and strategy guide communication, contributing to a genuinely pleasant atmosphere. Everybody, both women and men, has a voice without sarcasm. Tasks are distributed according to expertise and voluntarism. Power lies with the founding members, but agency is encouraged from all participants. The outcomes are the Helsinki initiative for NUP in 2016, signed by 7,000 people and the municipal council initiative in 2017, which led to the decision to start the formal feasibility study on founding the park. Currently, this is being implemented in collaboration with NUPH, which is also included in the new strategy of the city.

Gender is not salient in NUPH, although two-thirds of the activists are women. Tasks and roles are distributed on the basis of voluntarism and trust, but a handful of women mostly run the movement in practice. Ethnicity, sexual orientation and disability do not emerge as issues, but age does, as the members are mostly 50–60+.

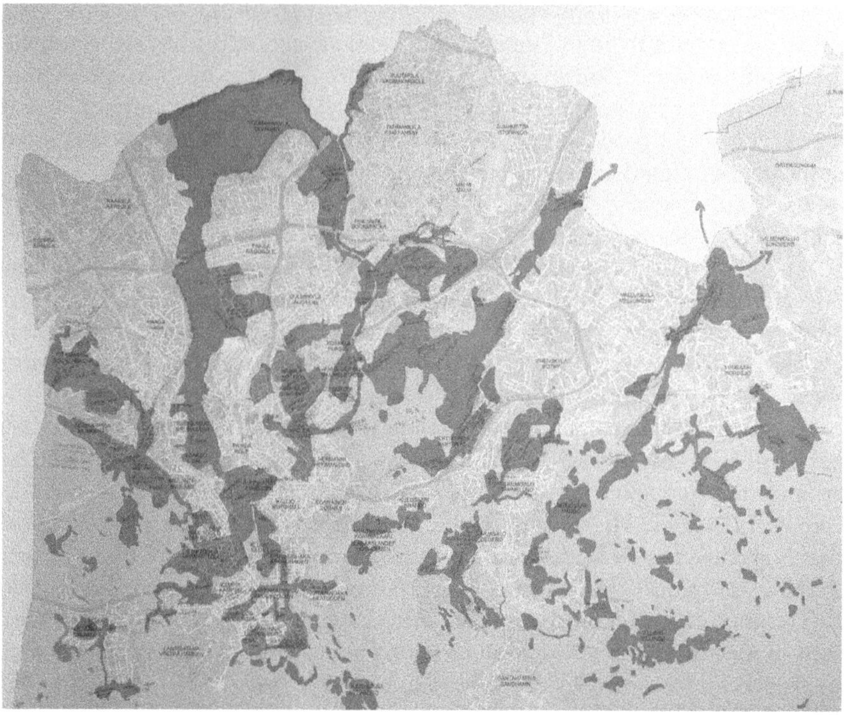

Figure 6.3 NUPH preliminary proposal – Helsinki as a nature metropolis – for the delineation of the park (permission by the City of Helsinki concerning the material of the feasibility study).[10]

Source: © City of Helsinki.

Comparison of the cases from the gender+ perspective

The aim of the comparison of the outcomes of the two cases is not to construct a hierarchical typology, but rather to contrast the different cases in order to better comprehend gender issues in (urban) planning and spatial development (Strauss and Corbin, 1990). The interpretation is based on the triangulation of the data-gathering methods (focus groups, observation, Webropol survey, analysis of archives) and content-analysis techniques (4R and Critical Frame Analysis).

There are many differences between the two cases, not only due to the size and temporal frames of the movement, but also due to the definitions of the problem and the suggested solutions (Table 6.2). The solution to the problem of MCH – a tight master plan – is seen as the main problem of the NUPH movement. However, the NUPH is not a problem to some of the members of MCH, as one of them points out. Nevertheless, the greatest discrepancy lies in the vision of what a good city is like. MCH has a narrow vision of what is good for

Table 6.2 Comparison of the two cases

Issues	*More City to Helsinki (MCH), founded in 2009, with 15,000 FB members*	*National Urban Park to Helsinki (NUPH), founded in 2015, 7,000 members + 80 neighbourhood associations*
The problem	Low density in the urban centre leading to a lack of affordable housing	The new effective master plan diminishing the Central Park
The solution	An effective master plan, leading to affordable dwellings in the city centre	NUPH, leading to a holistic mixture of natural and socio-cultural environments in support of all (ages, class)
Implementation	No explicit strategy but tactical urbanism; FB group, projects, events, networking	A positive, value-based strategy; networking and lobbying
Agency, voice and power	Informal triple hierarchy: FB moderators, core activists, followers. Voice is moderated. Male dominance. Women do not contribute to the strategic interests	Informal triple hierarchy: founding members, activists, followers. Both women and men have a voice and contribute to practical needs and strategic interests
Outcomes	Impact on the new master plan and Helsinki strategy. Urban planning has become a citizen issue	Impact on the Helsinki strategy. The formal feasibility study for founding the park has started in collaboration with NUPH
Doing gender+	The agent is a white, youngish (30–50), well-educated, cycling male. Women feel subordinated and deliver less (postings and other contributions), even if they have a more holistic view of the city. Ethnicity and disability not salient, nor sexual orientation	Gender neutral, although the majority are white, well-educated women (50–60+) who run the movement based on soft values. A holistic, multi-dimensional view of the city for all. Ethnicity and disability not salient, nor sexual orientation

the city. The female members in MCH have a more holistic vision than the MCH men, since their vision comprises children, aesthetics and environmental qualities, besides services and mobility, like in the vision of NUPH. Thus, women tend to stress the criteria "interplay of production and reproduction" and that of "private and public spheres" (see Table 6.1), whereas the men focus on "mobility and communication systems", in addition to dense "housing solutions". NUPH underlines the criteria related to the interplay of natural and landscaped settings.

Both movements are based on voluntary work and self-organisation with an informal triple hierarchy. However, MCH, or its communication in the FB group, is more controlled than behaviour in NUPH. The latter is guided by a holistic approach, which has contributed to experimental hybrid governance in the form of collaboration with the City on the feasibility study for founding the park. NUPH has proposed the following criteria for the choice of settings for the future park: nurturing cultural heritage, ecological sustainability, wellness and health, lifelong learning, culture and free-time, urban activism and economic well-being (see the criteria in Table 6.1 and Figure 6.3). The chosen settings might lead to an interesting urban landscape, when and if the politicians decide to implement the construction of the park. Irrespective of the different styles of implementation, both movements have been quite efficient in the sense that they have reached their goals – the new master plan and the decision to start the feasibility study on the park.

Similarities between the movements also exist in terms of their gender neutrality. Although 70 per cent of the members of NUPH are women, gender is not salient. The open responses in the Webropol survey disclosed that "gender has no significance" and that "the park is not a gender issue". Women and men participate equally. However, the evaluation discloses how gender+ is implicitly done. The NUPH women tend to run the business in practice and it was the women who were behind the choice of the values and ethical principles in the strategy.

In the male-dominated MCH women are clearly in a subordinate position regarding the power hierarchy, partly due to their communication style. On the other hand, gender does seem to play a role, because the women tend to have a more multi-dimensional and human view of the urban environment. They bring forth gender+ in terms of varying ages (children, elderly). The focus of the men,

Table 6.3 The dialectics between practical gender needs and strategic gender interests, based on individual and collective actions in Case 1, is male-dominated

Dialectics of:	*Practical needs*	*Strategic interests*
Individual actions	Participation in the FB group Participation in events and pub evenings	Densifying and expanding the city centre with new dwellings Negotiating projects with the officials Participation in political debates Using social media and other media to increase urbanity
Collective actions	Networking/Meshworking Moderation of the FB group Lobbying officials and politicians	Negotiation with politicians and officials Participation in hybrid governance via projects and the master plan proposal

in turn, is on the "cycling alpha-male". The other intersectional characteristics – ethnicity, class and sexual orientation – are restricted in the FB discussion.

The traditional critical question, posed by Maxine Molyneux (1985), concerns how to transform short-term practical gender needs into long-term, strategic interests. The examination of the individual and collective activities indicates that the dialectics between the practical needs and strategic interests is highly male-dominated in MCH (Figure 6.3). Women have a minor role in the FB group, in the events, projects and also in the impact on the final outcome. Table 6.4, in contrast, discloses that both women and men in NUPH have shared practical gender needs that have been successfully translated into strategic interests. They have been doing gender, not explicitly but in synergy, and this has brought forth additional value.

As evaluations most often end with recommendations, the following can be suggested on the basis of the comparison:

1 The double role of MCH, on the one hand as an overtly one-issue political movement and on the other hand as a general discussion forum for urban matters that attracts many people, is problematic. In addition, the rule of not allowing discussion of the vision of MCH and the narrow concept of urbanity and space in the FB group might be counterproductive, as they diminish the significance of the quality aspects of the city centre. The seminal question is how to integrate densification with quality. This is where gender might play a role, since in this case women seem to be the carriers of a more holistic view than the men, who are more focused on "bulk".

Table 6.4 The dialectics between practical gender needs and strategic gender interests, based on individual and collective actions in Case 2, is influenced by both women and men

Dialectics of:		
	Practical needs	*Strategic interests*
Individual actions	Finding facts for the Helsinki urban park narrative Contacting politicians and officials Participation in meetings and events	Saving and nurturing the ecological and socio-cultural heritage of Helsinki Collecting names for the municipal petition Participation in political debates
Collective actions	Networking/Meshworking Organising and coordinating the work in practice Organising events Participating in the feasibility study	Co-creation of a strategy, based on soft values Negotiation with politicians and officials Contributing to hybrid governance as a pilot

In addition, the stated rule in the FB group also prevents discussion of the structural prerequisites of housing production, for example the lack of metropolitan governance in the capital city region, regional politics in general and the development of other city centres, as well as the failing housing policy (Loikkanen and Laakso, 2016). There is also the question of whether the density politics of MCH will benefit not citizens in general, but instead investors and construction firms. This has been the case in London and San Francisco, where real estate developers have co-opted YIMBY movements and gentrified the city centres (Meronek and Szeto, 2017).

2 The NUPH movement, on the other hand, could learn from the effective application of tactical urbanism from MCH, especially in terms of moderating their FB group, although with softer rules. In turn, MCH could learn from the strategy and choice of values of NUPH, which have led the way to an experimental, transparent hybrid governance with the City. In addition, MCH, which does welcome women as members (applying the strategy of inclusion; Squires, 2005), could make an effort to empower women by engaging female moderators and by organising events with and for all genders.

3 Both movements could learn from the gender-sensitive literature on urban planning (Damyanovic, Reinwald and Weikmann, 2013; Sánchez de Madariaga and Roberts, 2013; Wankiewicz, 2016), since it provides tools for quality in spatial development, especially in terms of what urban density means from the user perspective.[11] The positive values and approach, as well as the smooth running of the movement in practice that can be attributed to the women in NUPH, should be marketed as an example of how to create a constructive atmosphere in which men can also flourish.

After the feedback of the results and recommendations to both cases, MCH did not change its rules concerning discussion about the vision, but it engaged a female moderator in the administration team. It also promised to coach other urban policy movements on how to create a powerful FB group, which NUPH is currently profiting from, as it seeks to improve its social media skills.

Discussion

The aim of the chapter was to present and discuss the methodological results of gendered evaluations of two cases in the context of self-organisation around spatial development in Helsinki. The core question concerned "how to monitor and evaluate spatial development and its impact from the gender+ perspective at the project level". We argued that the complexity of engendering evaluations of urban planning requires an integrated evaluation framework and the application of mixed methods in order to facilitate knowledge building and activate "learning to learn" behaviours in the management of urban issues. The conclusions in the discussion are based on the Quasi-Judicial (QJ) case method

(Bromley, 1986) in which claims are backed up by a network of empirical facts, context data and relevant concepts (warrants) from the framework.

We actually created, as we claimed, an integrative framework comprising concepts from gender studies, planning literature and feminist/gender evaluation approaches and applied a methodological package of mixed methods. So how did we fare, what did we learn and did it lead to evaluative thinking about gender in spatial development?

The context matters

The relatively new phenomenon of self-organising in cities, leading to urban pop-up culture and alternative economic activities impacting spatial development, has been mostly boosted by female participants, not only in Helsinki but also in Berlin, Amsterdam, Athens and Milan (Horelli, 2017). However, the phenomenon did not bring forth gender as an explicit issue in Helsinki. This might not have been the case if our evaluation examples had had clear gender objectives, but we could not find good documented examples. Therefore, it was the Finnish "gender order" which provided the norm for doing gender. Nevertheless, the evaluation methodology enabled us to dig deeper and to uncover layers of gendered patterns and the salience of doing or undoing gender (Deutsch, 2007).

Doing gender+ implicitly

As gender is a psycho-corporal and a socio-cultural process of doing gender+ in context (Butler, 2004), it has structural, procedural and symbolic consequences. According to Deutsch (2007), doing or undoing gender involves producing relations of dominance and submission in a specific socio-cultural, political and economic context or trying to undo them. The adopted methodology enabled us to disclose how gender was implicitly done in both cases. The men in the male-dominated MCH behaved according to the stereotypical male norm, as they were active in both fulfilling practical needs and transforming them into strategic ones (Figure 6.3). The women in the movement felt uneasy about the demanding communication style, but expressed – without aggressively defending – their wider view about the development goals for a good city. Consequently, the women were both horizontally and vertically segregated in terms of gender, which is, according to Yvonne Hirdmann (1996), one of the major mechanisms that maintains the unequal patterns of the patriarchal gender system (see also Escobar, 2017).

Nor was gender explicitly done in the women-led NUPH. On the other hand, both women and men in the movement found it natural that women run the show in practice (a norm in Finnish culture), which corroborates that women "can". Also the successful strategy, based on soft values and a holistic vision of the city, was credited to the founding women of the movement. Deutsch (2007) has claimed that doing gender varies in terms of salience.

However, the women's implicit gender consciousness, which contributed to the positive impact on the implementation of their solution and outcomes, could and perhaps should have been more salient. This would have served as a model for other women and men, even in the MCH movement. By contrast, the men's role in NUPH was somewhat transformative, as the men also chose to adopt the values based on a holistic vision and a partner-like style of communication. Could this be interpreted as indicating that action in NUPH was based on "the biology of love" (Escobar, 2017) or on a culture of appreciation that enabled gender divisions to be transcended? Nevertheless, hardly anybody did the + in gender! Thus the lack of occurrences of or references to ethnicity, class, disability and sexual orientation was deplorable but not surprising.

Self-organising around spatial development requires consciousness raising

The evaluation showed that self-organised urban citizen movements can impact decision-making and consequently the shaping of the environment, as was the case with both MCH and NUPH. It also disclosed that the two movements had different visions, ways of implementation and outcomes. The male-dominated MCH pushed for a dense urban habitat at the centre of the city regardless of environmental quality. The women-led NUPH strove for a holistic reproduction of the city with ecological and socio-cultural settings for all. Thus, self-organising is a "wild card", as it can have both positive and negative spatial impacts (Uitermark, 2015). The latter is especially true if there is no gender consciousness (Horelli, 2017). Therefore, consciousness raising and knowledge concerning the quality and justice of the urban environment is vital (Damyanovic, Reinwald and Weikmann, 2013; Sánchez de Madariaga and Roberts, 2013; Wankiewicz, 2016).

In these cases the women's more humanistic environmental visions and awareness of the interplay between reproduction and production as well as that between the public and private spheres were not due to essentialism, but to their experience of caring for both children and the environment. In addition, the more senior women (50–60+) in the NUPH movement still remember the human scale of the centre of Helsinki, in both their minds and their bodies. This is not an issue of YIMBY, but a site for transparent, democratic deliberation and co- or hybrid-governance with varying genders of different backgrounds.

Necessity for an integrated framework and mixed methods

The integrated theoretical framework comprised concepts that could be applied in the analyses and interpretations of the data and results. Methodologically it was a combination of regarding gender both as a variable (number of women and men) and as a construction and action (doing gender in processes; see Horelli, 2017). Also the application of mixed methods (triangulation), especially the 4R method (JämStöd, 2007) and Critical Frame Analysis with sensitising questions (Verloo, 2007), enabled the processing of the findings. This would not have been possible with mainstream manuals on gendered evaluations (Segone, 2015), which work well in developing countries where the ability to evaluate the programmes and projects in terms of gender is high. However, the two evaluation cases were not commissioned by the movements, which meant that the purpose and terms of the evaluation were not as clear as they could have been.

This comparative evaluation was carried out at the project level, meaning that it did not deal with programmes or policies concerning spatial development, which might have more transformative consequences. Engendering these types of evaluations would also require an integrative theoretical framework and mixed methods but, in addition, an expanded approach both to spatial development and gender (Horelli, 2017). There is evidence that self-organisation around spatial development and place-based globalism (Harcourt and Escobar, 2006) will increase in the future, due to digitalisation and social media. It will then be an essential element in (urban) complexity management (Boonstra and Boelens, 2011; Wallin, 2019), which in turn will greatly challenge the engendering of these new approaches to spatial development and their evaluation. This hopefully implies that the adopted gender perspective will successfully contest the systemic level, including the dominant patriarchal worldview that subjugates both women and men, contributing to the paradigm of solidarity.

Notes

1 We use "engendering" as a generic term for integrating equity/equality in urban planning, and "gender mainstreaming" as a formal, top-down way to incorporate equality in policy, programmes and projects.

2 Personal interview with architect Eva Kail (see also Brisolara, Seigart and Sengupta, 2014).

3 Commoning means engaged action in building, managing and nurturing shared resources for successful "commons", whether material, spiritual or socio-cultural (Bollier, 2014).

4 Gender neutral in the sense that environmental issues or those of everyday life do not get politicised as having a gendered relationship, such as the gender pay gap, gendered labour and education markets. However, the #MeToo campaign has raised awareness in Finland, especially of violence and sexual harassment against women.

5 The publication "Manual for Gender Mainstreaming in Urban Planning and Urban Development" by Damyanovic, Reinwald and Weikmann (2013) includes detailed descriptions of how to integrate the gender perspective in master, land-use and public spatial planning, as well as instructions on how to deal with housing construction and public service buildings from the gender+ perspective.

6 The 4R method is developed by the Swedish Gender Mainstreaming Support Committee, JämStöd from the original 3R method by Gertrud Åström. It provides information concerning R1 (gender representation in activities and decision-making), R2 (resource allocation between women and men), R3 (realia, i.e. the extent to which representation and resource distribution produce gender patterns which have been affected by gender norms in the organisation or in the setting) and R4 (the realisation of new goals and measures to achieve them).

7 Frames refer to the central organising ideas or interpretation schemes that structure the meaning of reality or its narrative account (Verloo, 2007).

8 Floor area ratio (FAR) is the ratio of a building's total floor area (gross floor area) to the size of the piece of land upon which it is built. The current FAR in the centre of the city ranges from 1.50 to 4.50. The new neighbourhoods of the city centre range from 2 to even 16 FAR.
9 The Ministry of the Environment has defined the quality criteria for NUP. They comprise important natural settings enhancing biodiversity (forests, shore and water areas), green areas and environments that are significant for the nation's history or the city's own development and social life. The ensemble should be sufficiently large, integrated and ecologically continuous.
10 From the 260 drawings by the NUPH participants it is possible to identify a few concentrations that emerge at the waterfronts, archipelago, sea-areas, forests and a cohesive greenbelt.
11 The application of only one density indicator, such as FAR (floor area ratio), does not provide a sufficient measure for the experience of the setting, but should be complimented by number of floors, dwellings/ha, green area space, etc.

References

BetterEvaluation (2017). *Evaluation standards*. [Online]. Available at: www.betterevaluation.org/en/evaluation-options/evaluation_standards [Accessed 28 Aug. 2017].

Bollier, D. (2014). *The commons as a template for transformation*. [Online]. Available at: www.greattransition.org/document/the-commons-as-a-template-for-transformation [Accessed 1 Dec. 2017].

Boonstra, B. and Boelens, L. (2011). Self-organization in urban development. Towards a new perspective on spatial planning. *Urban Research Practice*, 4, 99–122.

Brisolara, S., Seigart, D. and Sengupta, S. (Eds.) (2014). *Feminist evaluation and research. Theory and practice*. New York: The Guildford Press.

Bromley, D.B. (1986). *The case study method in psychology and related disciplines*. Chichester: John Wiley and Sons.

Bustelo, M. (2017). Evaluation from a gender+ perspective as a key element for (re)gendering the policymaking process. *Journal of Women, Politics and Policy*, 38(1), 84–101.

Butler, J. (2004). *Undoing gender*. New York: Routledge.

Chen, H.-T. (2005). *Practical program evaluation. Assessing and improving planning, implementation and effectiveness*. London: Sage.

Damyanovic, D. (2013). Gender mainstreaming as a strategy for sustainable urban planning. In: I. Sánchez de Madariaga and M. Roberts (Eds.), *Fair shared cities. The impact of gender planning in Europe*. Farnham: Ashgate, 177–191.

Damyanovic, D., Reinwald, F. and Weikmann, A. (2013). *Manual for gender mainstreaming in urban planning and urban development*. Vienna: Department Urban Development and Planning.

Deutsch, F.M. (2007). Undoing gender. *Gender and Society*, 21(1), 106–127.

Escobar, A. (2017). *Designs for the pluriverse. Radical interdependence, autonomy, and the making of worlds*. Durham, NC: Duke University Press.

Fleury-Bahi, G., Pol, E. and Navarro, O. (Eds.) (2017). *Handbook of environmental psychology and quality of life research*. Cham: Springer.

Goffman, E. (1974). *Frame analysis. An essay on the organization of experience*. Cambridge, MA: Harvard University Press.

Harcourt, W. and Escobar, A. (Eds.) (2006). *Women and the politics of place*. Bloomfield, CT: Kumarian Press.

Hillier, J. and Healey, P. (2008). *Contemporary movements in planning theory*. Critical essays in planning theory, 3. Aldershot: Ashgate.

Hirdmann, Y. (1996). *Key concepts in feminist theory. Analysing gender and welfare.* Aalborg: Aalborg University.

Horelli, L. (2017). Engendering urban planning in different contexts – successes, constraints and consequences. *European Planning Studies*, 25(10), 1779–1796.

Horelli, L. (Ed.) (2013). *New approaches to urban planning. Insights from participatory communities*. Helsinki: Aalto University.

Horelli, L. (2012). Evaluation of urban planning from the gender perspective. *Evaluation Connections*, September, 4–5.

Horelli, L. (2009). Network evaluation from the everyday life perspective. A tool for capacity-building and voice. *Evaluation*, 15(2), 205–223.

Horelli, L., Saad-Suloinen, J., Wallin, S. and Botero, A. (2015). When self-organization intersects with urban planning. Two cases from Helsinki. *Planning Practice and Research*, 30(3), 286–302.

Horelli, L. and Wallin, S. (2010). The future-making assessment approach as a tool for e-planning and community development – the case of ubiquitous Helsinki. In: C.N. Silva (Ed.), *Handbook of research on e-planning. ICTs for urban development and monitoring*. Hershey: IGI Global, 58–79.

JämStöd (2007). *Gender mainstreaming manual. A book of practical methods from the Swedish Gender Mainstreaming Support Committee (JämStöd)*. Stockholm: Edita.

Laakso, N. (2015). *Pingviinin arvoinen suoritus. Sosiaalisessa mediassa tapahtuva osallistuminen ja kaupungin kehittäminen. Case Helsingin kaupunkisuunnittelu ja – kehittäminen Facebookin Lisää kaupunkia Helsinkiin ryhmässä (Performance worth a penguin. Participation in urban planning and development through social media in the Facebook group of Case MCH)*. Helsinki: Aalto Yliopisto.

Laaksonen, S.-M. and Porttikivi, M. (2016). *Constituting a partial organization in Facebook through metaconversation*. An unpublished article presented 1 November 2016 at the panel in Laituri, organised by the Society of Urban Planning and Urban Activism as a Resource-project.

Libreria delle donne di Milano (Ed.) (1996). *Das Patriarchat ist zu Ende. Es ist passiert – nicht aus Zufall (Patriarchy is no more. It has happened – and not by chance)*. Rüsselsheim: Göttert Verlag.

Libreria delle donne di Milano (Ed.) (1989). *Wie weibliche Freiheit entsteht. Eine neue politische Praxis (Sexual differences. A theory of social-symbolic practice)*. Berlin: Orlanda-Frauenverlag.

Loikkanen, H.A. and Laakso, S. (2016). *Tiivistyvä kaupunkikehitys. Tuottavuuden ja hyvinvoinnin kasvun perusta (Densified urban development. Basis of productivity and wellness)*. Helsinki: Helsinki Center of Economic Research.

Lydon, M. and Garcia, A. (2015). *Tactical urbanism. Short-term action for long-term change*. Washington, DC: The Street Plans Collaborative Inc.

Mäenpää, P. and Faehnle, M. (2017a). Civic activism as a resource for cities. *Helsinki Quarterly*, 1, 68–81.

Mäenpää, P. and Faehnle, M. (2017b). The fourth sector is reconstructing the society. Yimbycon blog post 5, yimbycon.com. *Evaluation Connections*, September, 6–7.

Matthaei, J. (2018). *Feminism and revolution. Looking back, looking ahead.* Essay at the Great Transition Initiative. [Online]. Available at: http://greattransition.org/ [Accessed 10 Aug. 2017].

Meronek, T. and Szeto, A. (2017). *YIMBYs. The darlings of the real estate industry*. [Online]. Available at: www.truth-out.org/news/item/40509-yimbys-the-alt-right-darlings-of-the-real-estate-industry [Accessed 8 Aug. 2017].

Mies, M. (1994). Frauenbewegung und 15 Jahre Methodische Postulate der Frauenforschung (Women's movement and 15 years of the methodological postulates of the women's research). In: A. Dietzinger, H. Kitzer and I. Anker (Eds.), *Erfahrung mit Methode. Wege sozialwissenschaftlicher Frauenforschung (Methodological experiences. Paths of the sociological women's research)*. Freiburg im Breisgau: Core, 31–68.

Molyneux, M. (1985). Mobilisation without emancipation? Women's interests, the state and revolution in Nicaragua. *Feminist Studies*, 11(2), 227–254.

Nadin, V. and Stead, D. (2008). European spatial planning systems, social models and learning. *disP – The Planning Review*, 44(172), 35–47.

Podems, D. (2012). *Feminist evaluation and gender approaches. An introduction to exploring the differences*. [Online]. Available at http://europeanevaluation.org/sites/default/files/ees_newsletter/2012-10_Connections.pdf [Accessed 15 Jan. 2018].

Puerari, E. (2015). *Urban public services innovation. Exploring the 3P and 4P models*. Dissertation. Milano: Politecnico di Milano.

Reimer, M., Getimis, P. and Blotevogel, H.H. (Eds.) (2014). *Spatial planning systems and practices in Europe. A comparative perspective on continuity and changes*. New York: Routledge.

Sánchez de Madariaga, I. and Roberts, M. (Eds.) (2013). *Fair shared cities. The impact of gender planning in Europe*. Farnham: Ashgate.

Segone, M. (2015). *National evaluation policies for sustainable and equitable development. How to integrate gender equality and social equity in national evaluation policies and systems*. New York: United Nations.

Sielbeck-Bowen, K., Brisolara, S., Siegart, D., Tischler, C. and Whitmore, E. (2002). Exploring feminist evaluation. The ground from which we rise. *New Directions for Evaluation*, 96, 3–8.

Squires, J. (2005). Is mainstreaming transformative? Theorising mainstreaming in the context of diversity and deliberation. *Social Politics*, 12(3), 366–388.

Strauss, A. and Corbin, J. (1990). *Basics of qualitative research. Grounded theory procedures and techniques*. London: Sage.

Uitermark, J. (2015). Longing for Wikipedia. The study and politics of self-organisation. *Urban Studies*, 52(13), 2301–2312.

Verloo, M. (Ed.) (2007). *Multiple meanings of gender equality. A critical frame analysis of gender policies in Europe*. Budapest, New York: Central European University Press.

Wallin, S. (2019). *Managing urban complexity. Participatory planning, self-organisation and co-production of space*. Dissertation. Espoo: Aalto University.

Wankiewicz, H. (2016). *Gender Planning – Gender Mainstreaming in der räumlichen Planung. Top-down- und Bottom-up-Strategien als Bausteine zu einer nutzerInnen- und gleichstellungsorientierten feministischen Raumplanung (Gender planning – Gender mainstreaming in spatial planning. Top-down and bottom-up strategies as key elements of a user- and equality-oriented feminist spatial planning)*. Dissertation. Salzburg: University of Salzburg.

Wotha, B. (2013). Urban governance and gender-aware planning. In: I. Sánchez de Madariaga and M. Roberts (Eds.), *Fair shared cities. The impact of gender planning in Europe*. Farnham: Ashgate, 137–157.

7 A new generation of gender mainstreaming in spatial and urban planning under the new international framework of policies for sustainable development

Inés Sánchez de Madariaga and Inés Novella Abril

Introduction

This chapter discusses how the new international agendas for sustainable development provide a framework within which a new generation of gender equality policies and tools should be advanced in spatial development. The new international agendas on sustainable development – most significantly the Agenda 2030 for Sustainable Development and the New Urban Agenda – have recognised explicitly and in important ways the need to address women and gender issues. This chapter argues that existing experience in the highly technical field of urban and land-use planning does not to date qualify properly as gender mainstreaming – understood as a systematic policy approach encompassing different levels and policy actions as defined by the United Nations Beijing World Women's Conference of 1995 (Council of Europe, 1998; Mazey, 2001). Existing experience in Europe and elsewhere in the field of urban and territorial planning is mostly drawn from individual projects, toolkits, participatory workshops with neighbourhood women; at best, from somewhat wider actions or programmes, although most of these are short lived. In order to fully implement the gender equality goals embedded in the Agenda 2030 for Sustainable Development and the New Urban Agenda, more effective gender mainstreaming approaches to spatial planning must be put in place, going beyond the experiences and practices we have seen so far in Europe.

This chapter looks first at the new international agendas on sustainable development, including the Agenda 2030, the Paris Agreements on climate change, the Sendai Framework for Disaster Risk Reduction and the New Urban Agenda (NUA). All of these agendas have integrated a solid recognition of gender – understood as equality between women and men – and women's issues as important dimensions for their successful implementation. The New Urban Agenda, unlike Agenda 2030, also mentions other groups with specific needs who can potentially be discriminated against, including the elderly, the young, refugees and migrants, and ethnic minorities. In several paragraphs of the New Urban Agenda some of these groups are listed together with women as potentially discriminated groups needing special attention. None of the international sustainability agendas explicitly address gender+ or intersectionality issues as considered in some chapters of this book.

These interconnected agendas provide an important framework and political backing for advancing such renewed, stronger, more structural and more technical gender policies and tools in urban and spatial planning across the world. Second, the chapter provides an argument about how far actual experience has gone in integrating gender into urban, land-use and territorial planning. To this aim it looks at the best recognised experience, that of the City of Vienna. The Vienna case provides a singular opportunity from which to explore key issues regarding urban planning and gender mainstreaming (see also Chapters 4 and 5). Third, the chapter presents a case study: The Basque Country's Regional Plan, DOT in its Spanish acronym (Directrices de Ordenación del Territorio), which suggests a possible model for other regional planning efforts in Spain and elsewhere. Gender mainstreaming in the DOT is a pioneering example of measures on gender equality, integrated within a large-scale regional planning endeavour with long-term validity, broad regional scope, binding elements for local government planning, diverse subject areas for intervention, and variety and breadth in the measures and actions it contains.

The chapter concludes by pointing out a number of criteria that should be an integral part of any efforts to put in place and implement more ambitious agendas for gender mainstreaming in spatial development, ones that are fit to meet the challenges set by the recently adopted international agendas for sustainable development.

The international agendas for sustainable development

2015 and 2016 were two key years that made a major breakthrough in the United Nations global agendas on sustainability: the 2030 Sustainable Development Agenda, the agreements on climate change, the Sendai Framework for Disaster Risk Reduction and the New Urban Agenda (NUA). These four important international agendas should create a foundation on which to base the aspiration to develop a new generation of more effective and systematic gender-based policies and tools in urban planning and spatial development. Also important and recognised by the NUA is the Addis Ababa Action Agenda on Financing for Development.

The Agenda 2030 for Sustainable Development

The first, Agenda 2030, involved the adoption in September 2015 of a new framework for sustainability on the planet, through the new Sustainable Development Goals, SDGs, replacing the Millennium Development Goals, whose mandate came to an end that year (United Nations, 2015c). The 17 SDGs and their 169 objectives were adopted at a summit of the Heads of State of all countries and drafted in an unprecedented multi-party process of deliberation involving thousands of people worldwide. They provide a framework for action to eradicate poverty, promote economic development, protect the environment, and promote peace and good governance in the world. Goal 5 refers specifically to gender equality, as it aims at achieving gender equality and the empowerment of women and girls, while Goal 11 concerns urban development, aiming at making cities and human settlements safe, inclusive, resilient and sustainable (United Nations, 2015c). The two goals also reinforce each other: SDG5 refers to the city, especially public places and safety, and SDG11 mentions gender as an important issue for inclusion, safety and sustainability. Of the remaining 15 SDGs, virtually all have significant implications for both gender and cities, some of which have already been explicitly identified in the goals and objectives included in Agenda 2030, as can be seen in Table 7.1.

The major issues addressed by the SDGs are obviously intertwined, and therefore many goals and indicators within each highlight the interdependencies between them. Broadly speaking, Goals 1 to 4 could be defined as eminently social: education, eradication of hunger and poverty, health. Goals 6 to 12 include aspects that very generically could be referred to as "infrastructure-based": water and sanitation, energy, economic development, industry and innovation, reduction of regional inequalities, production and consumption. Goals 13 to 15 have more of an environmental character: action against climate change, life under water, life on earth; numbers 16 and 17 refer to issues of good governance, peace and security. In the development of the public policies that will derive from these goals, the interdependencies between all the SDGs will be more clearly shown. This will also expand and improve understandings of how and in what ways both gender and town planning are integral parts of the practical implementation of any public policy formally framed within almost all other SDGs.

The Agenda 2030 provides a reference point for most of the other international agendas, particularly the New Urban Agenda which has integrated all the SDGs as substantive topics for action at the urban level. In particular, and for the follow-up and review processes, the NUA takes a common stand with the Agenda 2030 by integrating within its reporting system all of the SDG indicators that can be tracked at the urban level. These represent up to 60 per cent of the 169 indicators of Agenda 2030, providing an important avenue for interconnectedness and synergies between both agendas. It is important to also note that both gender and cities have standalone SDGs in addition to their cross-cutting implications for practically all other SDGs.

The Agenda 2030 SDGs, like the New Urban Agenda, and unlike their respective predecessors, the Millennium Development Goals and Habitat I and II, take into account the global North as much as the global South. This expanded perspective not only provides new insights on the interlinkages between rich and poor regions of the world, it also provides new opportunities for exchange of experience which can be particularly relevant for women and

Table 7.1 Selection of gender and urban relevant targets and indicators for SDG5, SDG6 and SDG11. Note indicators need to be sex disaggregated (United Nations, 2015c)

Selection of objectives relevant to both gender and cities, by SDG	
5.1	End all forms of discrimination against all women and girls everywhere.
5.2	Eliminate all forms of violence against all women and girls in the public and private spheres, including trafficking and sexual and other types of exploitation.
5.4	Recognise and value unpaid care and domestic work through the provision of public services, infrastructure and social protection policies and the promotion of shared responsibility within the household and the family as nationally appropriate.
5.5	Recognise and value unpaid care and domestic work through the provision of public services, infrastructure and social protection policies and the promotion of shared responsibility within the household and the family as nationally appropriate. Ensure women's full and effective participation and equal opportunities for leadership at all levels of decision-making in political, economic and public life.
5.6	Undertake reforms to give women equal rights to economic resources, as well as access to ownership and control over land and other forms of property, financial services, inheritance and natural resources, in accordance with national laws.
6.1	By 2030, achieve universal and equitable access to safe and affordable drinking water for all.
6.2	By 2030, achieve access to adequate and equitable sanitation and hygiene for all and end open defecation, paying special attention to the needs of women and girls and those in vulnerable situations.
11.1	By 2030, ensure access for all to adequate, safe and affordable housing and basic services and upgrade slums.
11.2	By 2030, provide access to safe, affordable, accessible and sustainable transport systems for all, improving road safety, notably by expanding public transport, with special attention to the needs of those in vulnerable situations, women, children, persons with disabilities and older persons.
11.3	By 2030, enhance inclusive and sustainable urbanisation and capacity for participatory, integrated and sustainable human settlement planning and management in all countries.
11.5	By 2030, significantly reduce the number of deaths and the number of people affected and substantially decrease the economic losses relative to gross domestic product caused by disasters, including water-related disasters, with a focus on protecting the poor and people in vulnerable situations.
11.7	By 2030, provide universal access to safe, inclusive and accessible, green and public spaces, in particular for women and children, older persons and persons with disabilities.
11.8	By 2020, substantially increase the number of cities and human settlements adopting and implementing integrated policies and plans towards inclusion, resource efficiency, mitigation and adaptation to climate change, resilience to disasters, and develop and implement, in line with the Sendai Framework for Disaster Risk Reduction 2015–2030, holistic disaster risk management at all levels.

gender issues.[1] It will also contribute to the greater impact of both international policies in Europe, whose participation in these UN forums has been comparatively weak thus far.

The Paris Agreements on climate change

The agenda on climate change implied a new agreement by the governments of the world, gathered together in December 2015, this time at the Paris Summit, entitled COP 21.[2] The 196 members of the United Nations Convention on Climate Change (UNFCCC) adopted a universal agreement setting out the rules for a new regime of adaptation and mitigation for the post-2020 period (United Nations, 2015a). In the view of experts, this was the last chance to reach a global agreement that would limit the increase in the average temperature of the Earth's surface to below the dangerous threshold of 2°C, as compared with the pre-industrial period. The legally binding agreement reached in Paris at COP 21 commits the signatory countries to maintaining the global temperature increase substantially below 2°C; funds of nearly US$100 billion for developing countries as of 2020, and finally, a review every five years.

The fact that this agreement has been reached is important from a gender perspective because women are the majority of those who suffer the negative effects of climate change, especially in low- and middle-income countries (MacGregor, 2017). There is also evidence that women in high-income countries contribute less to the causes of climate change and are more in favour of promoting measures to reduce it (Buckingham and Le Masson, 2017). In addition to reasons of democracy and representation, these are some of the reasons why the presence of women in the bodies responsible for designing and implementing policies to combat climate change is important (Kronsell, 2017).

From the point of view of urban planning, the Paris COP 21 agreements highlight a fundamental sphere of action for urban policies, since metropolitan and urban transport, and heating in the residential sector are two key greenhouse gas-producing sectors, polluting the atmosphere (Chapman, 2007; Ranson, Morris and Kats-Rubin 2014). In fact, municipal and regional action in the transport sector will be key to implementing the Paris Agreement.

As has been argued elsewhere, it is essential that environmental sustainability objectives are not achieved at the expense of the objectives of equality, and in particular of gender equality objectives (Sánchez de Madariaga, 2013). Contrary to what one might think and is often stated, they are objectives that do not always go hand-in-hand with each other, one reinforcing the other, but sometimes can lead to contradiction.[3] An example from the transportation sector comes from policies restricting access to cars by low-income women in the US on the basis of the environmental impacts of such increases in car usage (Blumenberg, 2016). It is important to make sure that policies supporting a reduction of emissions of greenhouse gases into the atmosphere by restricting car use and/or promoting increased use of public transport are not implemented fundamentally at the expense of transportation options for the female population.[4]

Sendai Framework for Disaster Risk Reduction 2015–2030

The Sendai Framework for Disaster Risk Reduction of 2015 aims at adopting a concise, focused, forward-looking and action-oriented post-2015 framework for disaster risk reduction by building resilience, considering the experience gained, identifying modalities of cooperation based on commitments, and determining modalities for the periodic review of the implementation of a post-2015 framework for disaster risk reduction (United Nations, 2015b). While recognising progress in building resilience and reducing losses and damages, it calls for a more explicit focus on people and their health and livelihoods, and regular follow-up. Over the next 15 years, it aims to achieve a substantial reduction in disaster risk and loss of life, livelihood and health and the economic, physical, social, cultural and environmental assets of persons, businesses, communities and countries.

The Sendai Framework recognises the importance of gender dimensions as disasters affect men and women, and boys and girls, differently because of existing socio-economic conditions, cultural beliefs and traditional practices. Women are more likely to be disproportionately affected, including increased loss of livelihood, gender-based violence, and even loss of life during, and in the aftermath of, disasters. In particular the framework calls for:

> ... a gender, age, disability and cultural perspective in all policies and practices ... the promotion of women and youth leaderships.... women and their participation are critical to effectively managing disaster risk and designing, resourcing and implementing gender-sensitive disaster risk reduction policies, plans and programmes. Adequate capacity building measures need to be taken to empower women for preparedness as well as build their capacity for alternate livelihood means in post-disaster situations.
>
> (United Nations, 2015b: 13)

The New Urban Agenda (NUA)

Within the framework of Habitat III, the Third Conference on Housing and Sustainable Urban Development, in October 2016 the UN member states adopted a New Urban Agenda entitled the "Declaration of Quito on Sustainable Cities and Towns for All", which lists a number of shared guidelines for urbanisation processes on the planet over the coming decades (United Nations, 2016). The NUA will guide the actions of a broad range of players in cities over the next 20 years and fully take on board the principles and objectives of the 2030 Agenda for Sustainable Development.

The NUA includes a significant number of new developments, compared to what was agreed in Istanbul in Habitat II, such as integrating urban processes in high-income countries, the reference to human rights in the city, the consideration of urbanisation processes as opportunities rather than problems, the importance of town planning tools, the effective means of implementation, and monitoring, follow-up and review processes and indicators. As a new feature, in its articles the NUA includes a total of 34 references to the terms "woman/en" or "gender" out of 175 paragraphs.

The preparation process for the NUA from September 2014 to July 2016 was an extremely complex process involving an extensive series of official and semi-official events, including four regional meetings (regions as understood at the global level, following the United Nations concept, meaning Europe is one of these regions) and seven theme-based meetings, from which statements were made. Particularly important were the World Urban Forum held every two years, the World Urban Campaign, and the many Urban Thinkers Campuses held all over the world, which provided discussion forums for the various unofficial players (Sánchez de Madariaga and Boccia, 2016).

Stakeholders representing different sectors of civil society were organised in various ways, with the General Assembly of Partners, GAP, becoming the main platform for multi-stakeholder participation, both during the process and after adoption of the NUA. The General Assembly of Partners is organised in 16 groups, representing such diverse constituencies as women, academia, local governments, professionals, older persons, persons with disabilities, youth, parliamentarians, farmers, etc.[5] Groups of experts organised into "policy units" prepared a series of recommendations for the drafting and implementation of the NUA, which became "issue papers" open for public consultation around the world. A Group of Experts on Gender, convened by UN-Habitat and the Houairou Commission, met at the Rockefeller Foundation in New York in 2015 and produced a document with detailed recommendations (Houairou Commission and Cities Alliance, 2016). The Houairou Commission, UN-Women, and other civil society actors such as the Cities Alliance, produced additional documents addressing specific gender issues throughout the negotiation processes until the very last minute.

In May 2016, the Zero Draft was published, based on all these materials. The first review was published in June; successive drafts, compilations of responses by member states, proposals by representatives of civil society and reviews were published during the month of July; public hearings with member states and additional consultation with stakeholders took place in New York in June. On 10 September the final text was approved in New York and this was formally adopted in October in Quito.

The First Quadrennial Report on the follow-up to and review of the implementation of the New Urban Agenda (United Nations, 2018) positions the New Urban Agenda vis-à-vis the other global development agendas adopted in 2015 and 2016, clarifying the interlinkages, describing how the New Urban Agenda is an accelerator for the achievement of all other agendas, and calling for coherence and integration in both policy and implementation. Section III recommends an incremental approach to reporting based on four elements: systems for the production of user-friendly and participatory data platforms that add to existing sources of data; new and expanded platforms for engagement on knowledge production and reporting; a United Nations system-wide coordination mechanism that

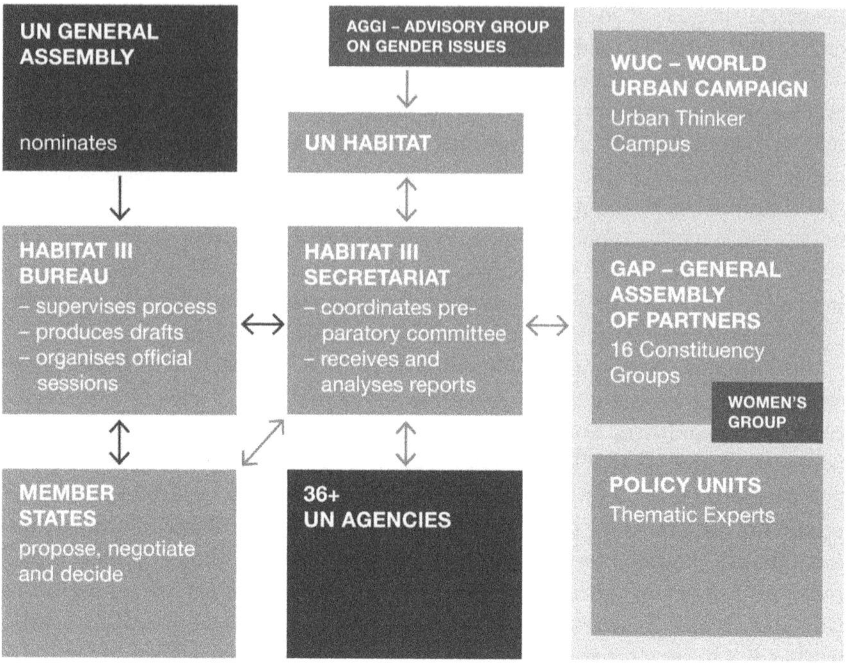

Figure 7.1 Flowchart of organisations and processes leading to Habitat III and the adoption of the New Urban Agenda.

Source: © Sánchez de Madariaga.

strengthens and expands partnerships; and capacities to report on the implementation of the sustainable urban development agendas, including data collection and analysis, policy development, implementation and stakeholder engagement (United Nations, 2018). The document further describes key stepping stones towards effective implementation, including governance structures and supportive frameworks; planning and management of urban and territorial spatial development; and means of implementation, including financing, innovations and capacity development.

Gender and the role of women in the NUA

Throughout this process there has been very active participation of experts on gender and, above all, grassroots women's associations from all over the world organised through the Houairou Commission, who have been at all the meetings and contributed proposals, ideas and recommendations for every text published successively and on all the declarations. Women have been extremely active in the various forums set up by the General Secretariat for Habitat III: The World Urban Campaign, the General Assembly of Partners and the policy units. Undoubtedly, women have been key protagonists who have been the most successful advocates from among all representatives of civil society. They were able to make their voice heard in the process in the most articulate and decisive way.

Judging by the high number of mentions of women and gender, the document finally adopted could be considered a success of women. Especially relevant is the mention in the introduction that sets a high standard:

> By reorienting the way in which cities and human settlements are planned, designed, financed, developed, administrated and run, the New Urban Agenda will … achieve gender equality and the empowerment of all women and girls in order to make full use of their contribution, vital to sustainable development…
>
> (United Nations, 2016: 4)

Other noteworthy mentions refer to: the representation of women on the public agenda and in urban decision-making processes; the economic empowerment of women; decent employment; unpaid domestic work and women in the informal economy; safe cities for women and safe public places; ensuring women's access to land tenure and housing. As a result of the historical developments of the New Urban Agenda since Habitat I and Habitat II, in which high-income countries were not part of the picture, these topics and the wording might seem and probably are closer to the realities of cities in developing countries. However, the wording is general enough to sustain action also in Europe, North America and other high-income parts of the world. Another criticism could arise from more demanding perspectives such as that expressed by, among others, Caroline Moser.

If we consider a strong definition of gender transformation as necessarily implying structural changes in gender power relations, a truly transformational inclusion of gender in the NUA would imply that the documents should somehow contain content that would enable a paradigm shift in city planning policies in order to focus on women in cities. Arguing for such a strong understanding of gender transformation, Moser (2006) links the transformation of gender to the accumulation of assets, including financial, social, natural and human. Assets create capacity for action and empower individuals and communities, and exist within social processes, structures and power relations. They enable transformations and, through collective actions, successfully challenge power relations. Applying Moser's criteria, only 3 of the 34 references in the NUA would fall under the category of transforming intervention.

If we apply this demanding criterion, the result does not seem to live up to expectations. Most references to gender are very generic and limit themselves to including women or gender as just one more in a list of marginalised groups, without considering specificity, differences between women or intersectionality. In general terms, the underlying consideration is of women as vulnerable and excluded victims, and not as positive and transforming agents of change. One important missing element is that there is no specific mention in the key sections regarding implementation and mechanisms for monitoring, follow-up and accountability. But overall, the result should be considered a success. It is no small thing that a United Nations document, signed by national governments around the world, mentions women or gender in 33 of its 175 paragraphs. This is a good basis, in combination with SDG5 and SDG11, on which women's movements, academics specialising in gender, local and regional governments, and all other urban stakeholders can support renewed action that could lead to a new generation of approaches to effective gender mainstreaming in urban planning.

What has been achieved so far? Falling short of real gender mainstreaming in urban and territorial planning

The field of urban planning is emerging as one of the areas of public policy least permeable to the integration of gender equality policies. Where gender equality is addressed experiences are often limited to pilot projects, one-off measures, specific approaches, or are restricted to a particular topic. In other cases, they lack efficient means of implementation, or the involvement of planning departments in charge of developing and implementing plans and regulations at local or regional levels. Programmes are often developed by women's or gender equality offices with little collaboration or influence in planning departments. On other occasions they have been of limited duration, their existence depending on the presence of specific individuals in decision-making positions who have driven them forward and implemented them during the term – whether longer or shorter – of their public office.

This assertion is based on the evidence by a number of recent publications providing a reasonable overview of the state of the art of actual practice. Within the European context, this evidence is made available by the book *Fair Shared Cities. The Impact of Gender Planning in Europe* (Sánchez de Madariaga and Roberts, 2013), which describes the main experiences up to that date in Europe.

At the global level, *Gender and Urban Planning* (Reeves, Parfitt and Archer, 2012) prepared for UN-Habitat, and the work by Jarvis, Cloke and Kantor in their book *Cities and Gender* of 2009, show sufficient evidence on which to ground this assertion for the developing countries. Beyond European, North American, and a few other countries, we must keep in mind that planning systems are often in their infancy and/or are mostly ineffective, lacking the means to guarantee effective compliance to those regulations that exist or to mobilise sufficient resources both financial and technical. These countries have now the chance, within the processes leading up to the implementation of the different international agendas and particularly the NUA, to integrate gender approaches as they improve and consolidate their planning systems.

Discussion of gender mainstreaming in urban planning in the developing world is, for these reasons, a rather different thing. However, it generally also involves a piecemeal approach rather than the substantive gender mainstreaming of urban policies. Gender initiatives at the urban level in these countries are frequently realised from the perspective of sectoral ministries, such as education, health, employment or others. Often they are implemented by national level administrations, as local governments in many countries are weak in fiscal, budget, staff and technical capacity. The issues most commonly addressed involve land-tenure and inheritance rights for women, promotion of women in the formal and the informal economy, gender budgeting, programmes to improve safety for women in urban space, toilet building, reproductive and other health issues, water infrastructure provision and access to energy for household cooking and lighting. Often these actions are put in place through projects and programmes of international cooperation managed through national level agencies, and are informed by the gender approaches developed over the last decades within the international development community.

Issues to explore through the case of Vienna – the challenging move from positive action into gender mainstreaming[6]

In high-income countries, Vienna, perhaps, is the exception. The Vienna case, which is presented in more detail in Chapters 4 and 5 of this book, raises many interesting issues after over 25 years of experience in gender equality projects and policies in planning and construction. Vienna implemented its gender planning policies mainly through the creation of a new high-level technical office. The full name of this Office created in 1998 was "Coordination Office for Planning and Construction geared to the Requirements of Daily Life and the Specific Needs of Women" (for the sake of simplicity, we will refer to it as Coordination Office in the following paragraphs). The Coordination Office was part of the Executive Group for Construction and Technology, to which 20 technical departments of the City report, a sub-organisation of the Chief Executive Office of the City. As a matter of fact, the Coordination Office was a "spin-off" of the Department of Women's Affairs, the so-called Women's Office, created earlier in 1992, which was also directed first by the same person, Eva Kail, an urban planner by professional background.

For many years, the activity of this Coordination Office consisted in implementing pilot projects in many different areas of planning and construction addressing the specific needs of women and girls in their daily living. The numerous and successful pilot projects run by this Office over more than a decade (1998–2009)[7] managed at the time to impact upon the regular and routine practices of the City. They achieved a not to be underestimated success in creating structural changes in municipal urban development which were notable in housing policy, transportation and park planning. This was made possible by the political support of the different City Councillors for planning, transport and housing (although to varying degrees) and the technical leadership of a small number of people, mainly Kail and her two female co-workers Elisabeth Irschik and Claudia Prinz-Brandenburg. It can be posited that these impacts across departments, which involved significant and structural changes in routine operations of housing and transportation policy, should be recognised as the foundation of a gender mainstreaming process in planning issues.

As the authors state in Chapter 4 in this book, after 2000 gender mainstreaming became a cross-cutting strategy for the whole municipality including also in spatial planning. In 2005 a general Gender Mainstreaming Unit was installed, situated in the Chief Executive Office with a mandate of addressing gender issues within municipal policies and programmes across the board. The Unit's head, Ursula Bauer, had previously worked in the Women's Office until 2005. After these organisational changes, gender mainstreaming specifically within spatial planning was made effective at a first phase (2005–2010) through the requirement that all the departments connected to spatial planning had to develop gender pilot projects in close cooperation with the Coordination Office. Such projects were included as part of their annual contract within the frame of a new public management organisational system between the department and the City Councillor of planning and transport, who was very supportive of gender planning issues. This first phase resulted in about 60 projects of various dimensions.

A second phase involved a more structured implementation involving the development of toolkits (Damyanovic, Reinwald and Weikmann, 2013), manuals, guidelines, knowledge transfer, evaluation, and reporting procedures. In addition to this, successive Urban Development Plans for the City of Vienna make increasingly relevant references to women and gender, including those issued in 1994, 2005 and 2014. This second phase, aiming to achieve a more systematic approach to gender mainstreaming, coincided in time with the resolution of the high-level Coordination Office responsible for gender in planning and construction policies, following the arrival of a new female director at the Executive Group for Construction and Technology in 2010.

The new director "mainstreamed" the gender planning experts: Eva Kail was transferred within the Executive Office to the Central Coordination Unit for Planning and Urban Development as a co-worker of the planning director. The responsibilities of this Unit for Planning include overall urban planning, the Smart City Strategy, participation processes, along with gender planning. Since then, a new pilot project on urban design and gender-sensitive participation has been realised (Chapter 4) and a gender planning manual has been published. The two other gender planning experts now work in the Department for Architecture and Urban Design in the units for public space and educational buildings. The cooperation of Eva Kail with the Gender Mainstreaming Unit of the City involves mainly gender training and the assessment of strategical planning documents.

In short, these organisational changes have resulted in a reduction of resources from a strategic unit addressing specifically gender issues in spatial planning to a single gender planning co-worker in the strategic unit for mainstream planning, and, very importantly, in a loss of visibility. In the Viennese case, the question is not so much the sustainability that specific initiatives could have. In fact, gender and women oriented policies have proved extremely resilient to political and institutional changes. The stable network of gender experts within the City is surely an important positive factor. Rather the issue is how the structural transformation of a number of planning practices of the City – which undoubtedly resulted from the many pilot projects and through the use of the numerous manuals and work aids during and after the Coordination Office's period of existence – will transform into a further consequent roll-out of a full gender mainstreaming policy, without a specific office responsible for gender policy within spatial planning.

It is often the case with gender policies that when the department in charge of creating a policy specifically addressing women's or gender issues disappears – a step that is often supported by the argument that "we are now doing gender mainstreaming" – the gender policy is de facto diluted. We must bear in mind that cross-cutting gender within relevant departments must be added to specific action directed at women by a specifically designated unit, rather than replacing it. Both reinforce and need each other.[8] Mainstreaming understood in the narrow sense as "cross-cutting" does not replace the existence of specific measures designed and implemented by an ad hoc administrative body. It supplements it.

Replacing positive action and ad hoc departments responsible for practical implementation, when mainstreaming gender policies are decided on, often result in the disappearance of an effective equality policy. Similarly, the implementation of a cross-cutting gender policy where there has been no previous positive action targeting women requires the parallel creation of ad hoc bodies and positive action measures. Gender mainstreaming in urban planning, as in any other field of public policy, must include both cross-cutting measures and positive action measures and a corresponding administrative framework responsible for their definition and implementation.

In the case of Vienna, the actual situation is more complex as it still provides gender capacities, and therefore a long-term development is difficult to judge. There is the Vienna women's affairs department, addressing women's and girls' interests, and the general gender mainstreaming unit and the gender planning expert in the Chief Executive's Office. But in fact the explicit personnel resources for coordinating and stimulating gender planning were reduced parallel to a shift in the political attention, and therefore the explicit gender planning activities also declined whereas mainstream activities were influenced to a certain degree. Recently more explicit gender planning activities are taking place again as a result of resumed political attention.

Although organisational culture and planning frameworks are not easily comparable, it would be illuminating and important for the efforts that will undoubtedly be made in the coming years in many cities to have more systemic studies of the Vienna case – from a cautious time-based distance – that could highlight some lessons for all.

A pioneering experience at the regional level: the Basque Country's Regional Plan (DOT)

The gender mainstreaming approach in the revised Basque Country's Regional Plan (DOT, Directrices de Ordenación Territorial) represents a pioneering experience in Spain and Europe on how to develop a broad programme for gender mainstreaming within a long-term planning policy, addressing regional-level spatial planning and the local planning decisions of municipalities (Sánchez de Madariaga, 2016).[9] The DOT are the higher level of planning in the Basque Country on which Partial Territorial Plans and Sectoral Regional Plans are based. The DOT have the following purposes: to build a framework for sectoral policies and for municipal planning; to formulate a set of principles, norms and criteria guiding and regulating urban development processes in order to guarantee territorial balance; and to guide the territorial actions requiring joint action by the Spanish government, the Basque government and/or other regional governments. In particular the DOT establish a series of regulations covering a wide scope of aspects including natural and agricultural space, areas to be developed through Partial Territorial Plans, areas to be set aside for large infrastructures and facilities, the quantification of residential space, the global estimate of historic preservation of architectural heritage, and the definition of the causes and situations under which the DOT should be modified or adapted.

The DOT were adopted for the first time in 1997. The document recently approved, after a three-year participatory process involving citizens and a broad range of stakeholders, is technically a revision of the one approved 20 years ago and is aligned with the Agenda 2030, the NUA and the Paris Agreements on climate change. In particular it includes climate change as a cross-cutting issue. Gender within the DOT is specifically addressed and also considered as a cross-cutting issue. Gender is understood as equality between women and men, taking into account intersecting issues such as age, socio-economic level, rural living and others.

The Basque Country is a region of 7,200 km^2 in Northern Spain with a population of 2,155,000, distributed in four cities over 100,000 inhabitants including the metropolitan area of Bilbao with 350,000, many other smaller towns, and a substantial landscape of rural villages and hamlets. With a regional autonomous government, the Basque country has the powers to adopt and implement its own legislation both in the field of urban regional planning and also in the field of gender equality. In both fields Basque legislation is noted for its advanced provisions. The 2005 Basque regional law for equality between women and men (Ley para la Igualdad de Mujeres y Hombres del País Vasco, Ley 4, 2005) broadly develops the concept of gender mainstreaming to be implemented across all fields of policy, stating in Article 3.4 that public governments must integrate gender perspectives in all their policies and actions with the objective of eliminating all kinds of inequality and of promoting equality between men and women. The law integrates a significant number of provisions, specifically referring to urban planning, urban amenities and housing policies.

In terms of gender policies, this has meant the creation over the years of positions for gender experts spread throughout public administrations at all levels and in all substantive areas of policy. For over 15 years local administrations in the Basque country, including the big cities of Bilbao, Vitoria and San Sebastián, have been implementing measures on safety in public space through participatory processes with women.[10] An early example was seen in San Sebastián in 1997, under the leadership of a Women's Platform called Plazandreok. Basque municipalities have created a Network of Basque Municipalities for Gender Equality.[11] Jáuregui (2012) provides an overview of initiatives at the local level on gender in planning in 88 municipalities of over 3,000 residents, representing 83 per cent of the population in the region. The results show 40 per cent of Basque local governments have implemented some kind of initiative on gender in planning.

The proposal for the DOT opted to make a sufficient, strategic and prioritised selection of measures, intentionally ruling out the option to systematically cover all fields of action in the DOT for the purposes of effectiveness. This strategic option was based on the previous experience of one co-author on gender mainstreaming European science policy as Director of the Women's Unit in the Cabinet of the Minister of Research and Innovation (Sánchez de Madariaga and Raudma, 2011). The experience at the European Commission of gender mainstreaming science policies demonstrated the risk of attempting to cover an overly broad spectrum of action, which could ultimately lead to a lesser impact (Abreu *et al.*, 2017). In pioneering actions on gender, as is the case here, it is important to ensure the applicability and success of what the administrations commit to do. An excessively broad or overly ambitious programme of action, difficult to implement, is hard to put into practice and the risks of failure are substantial.

Of the eight subject areas into which the DOT Baseline Document is structured, priority action was proposed in four, where the main aspects from a gender perspective are especially relevant: Urban environment and land-use planning, Rural environment, Sustainable mobility, Governance. Although there are significant gender dimensions in the other DOT subject areas – energy, water, landscape, physical environment – we chose not to propose specific measures in these areas, deciding rather to concentrate efforts on areas where gender dimensions are more relevant. In addition, considering that the various areas are interrelated, a number of the measures proposed for some of the selected areas have a knock-on effect on other areas. For example, measures in sustainable mobility affect the area of energy, which we have not selected as a priority area of action. The specific actions proposed in each of the selected priority areas are as follows:

Box 7.1 Summary of actions addressing gender topics proposed, by a selection of strategic areas of the Basque Country Regional Plan (Sánchez de Madariaga, 2016)

Urban environment and land-use planning

- To draft mandatory instructions for including land reserves in municipal land-use plans dedicated to providing facilities for the care of minors and older persons
- To develop a Care Facilities Plan for minors and older persons in collaboration with city councils
- To implement pilot projects through collaborations with city councils, through public call procedures
- To create financial incentives for gender-based action through co-financing by the Basque Government of municipal programmes, plans or projects
- To draw up guidelines and recommendations addressing technical staff for the introduction of a gender perspective in new urban developments and in urban renewal work
- To draft specific measures to be integrated within the Housing Plan
- To identify locations perceived as "unsafe" through participatory processes with women and other social groups (elderly persons, girls and boys)
- To draft guides for public place projects and funding from the gender perspective that will be applicable in new building projects promoted by the Basque Government or city councils

Sustainable mobility

- To review mobility statistics from a gender perspective to remove biases and better describe the mobility linked to care work
- To include the gender perspective within the new regional and urban Plans on Sustainable Mobility (PMUS) or in the review of existing plans, bearing in mind the concept of "mobility of care" (Sánchez de Madariaga, 2013)

Rural environment

- To improve the public transportation network and/or transportation services on demand in rural areas, linked to the Amenities Plan
- To highlight visibility of work done by women in the farming environment and their role in the rural environment, empowering them as farm owners

Governance

- To design processes of citizen-based participation with a gender perspective, aimed at boosting the involvement of women
- To draft programmes to support the professional careers and leadership of women in relevant professional sectors
- To review existing processes in government departments, at both self-governing region ("autonomous community") and municipal levels, aimed at drafting proposals to improve the inclusion of expert knowledge on gender at the most relevant stages of the decision-making process and its implementation

Elements for the spatial implementation of the gender dimensions of the sustainability agendas

The 2030 Agenda for Sustainable Development, the Paris Agreements on climate change, the Sendai Framework on Risk Reduction and the New Urban Agenda usher in a new international scenario, backed by national governments around the world, where gender equality and sustainable urban development are fundamental principles. These ambitious agendas provide a solid starting point from which to lay the foundations for a quantitative and qualitative leap forward in gender equality policies and tools in city planning. Learning from past experiences and the accumulated wealth of academic research can enable us to go beyond pilot projects, specific programmes or fairly short-term programmes that do not survive the move of the individuals who initially promoted them.

This new generation of gender mainstreaming should rely on both deliberative, participatory processes, and the planning powers of the state. The planning powers of the state translated into plans and regulations are particularly important in the highly regulated field of urban and land-use planning, a field which involves very significant financial public and private investments that are immobilised over the long term. The importance of the powers of the state in the field of planning has become evident in recent decades, also in liberal democracies and even under the processes of deregulation, privatisation, and the various forms of involvement of private sector and civil society stakeholders that we have seen in Western democracies since the 1980s (Alterman, 1988). Although unlike many public services, urban planning and land-use control cannot be significantly privatised, modern planning systems integrate ample mechanisms for the participation of grassroots citizens and private sector organisations. This allows for various modes of stakeholder involvement, public-private forms of collaborations, and citizen participation, embedded in each country's legal, political, administrative and planning contexts.

Plans are often implemented in ways that differ from what was initially intended. This occurs through mechanisms allowing for such modifications established in legislation and/or caselaw practice, managed through the various means of planning control established mostly at the local level. Rather than diminishing, the amount of administrative, technical, and legal inputs into planning is, if anything, expanding around the world, as new regulations regarding environmental, health hazards and other issues are increasingly put into place through permits and required reports, such as impact statements, frequently granted at regional or state levels. When private actors are powerful corporations, planning decisions can sometimes be overly influenced by private interests. This has happened to the point that the courts have on occasion intervened and declared an illegal use of mechanisms allowing for public-private partnerships.[12] In such cases, the courts have ruled that adopting such agreements implied de facto an illegal sale of public planning powers. Such legal rulings set the limits of how far public-private agreements can go in planning practice in one of the areas more susceptible to influence by powerful private actors, that is, the mechanisms specifically designed to allow public-private partnerships in urban development projects of a certain scale, to be developed over a number of years, in which the roles, responsibilities and financial commitments of both sides are carefully specified in legal contracts.

Bottom-up processes, however, are not those in which the private stakeholders are powerful developers or financial institutions, but instead neighbours' associations[13] and professionals acting in a role akin to what Davidoff (1965) referred to as *advocacy planning*. In most developed countries planning today involves many more such bottom-up approaches than in the 1960s. However, we would like to point out several issues with respect to the scope and nature of such approaches in the urban and land-use planning area. Such bottom-up activity is normally an integral part of formal planning systems in terms of how it is actually carried out. It is very often also funded by public administrations. For example, the participatory workshops and walks with women to identify unsafe locations carried out in Catalonia are funded by local and regional governments. Exceptionally, bottom-up processes are neither of these two things: they are not funded by the public administrations, nor are they part of the formal planning systems. One such rare occasion occurred in the renewal of slum areas in the city of Madrid in the aftermath of Franco's dictatorship in the late 1970s and early 1980s. In those years of political turmoil neighbours' associations backed by advocate planners and architects were able to obtain legal rights and titles to the property on which their slums had been illegally built, and also to have the local government bring in all kinds of infrastructure. This was an extraordinary example of a successful, entirely bottom-up process. But even in such cases, at the end of the day, in order to have the built environment transformed, the successful vindications of the slum dwellers, obtained through hard-fought political battles, had to be translated into technical and administrative documents following legal planning procedures. Whether we like it or not, all things relevant to the physical transformation of the built environment in modern societies ultimately have to go through legal administrative procedures and be approved by the competent public administration.

So a first underlying assumption involves the notion of what urban and spatial planning *is* and what it *is not*. Urban and spatial planning is a complex system involving institutions, regulations, laws, technical tools and means for effective implementation, public budgets and investments, mechanisms allowing for public-private agreements and scope for bottom-up citizen participatory processes, specialised civil servants with higher education degrees on the subject, working in sufficiently staffed and resourced planning departments, having the mandate to define the plans and manage their implementation. It uses the powers of government, including that of assigning use to particular plots of land, and of eminent domain, to decide and act upon the location and characteristics of infrastructure, housing, facilities, offices, and the like, as well as the preservation of open space.

Of course, planning involves private organisations and civil society groups organised in various ways and through different kinds of participation, sometimes formally integrated in the systems, at other times proceeding in more open ways. Certainly, it can be made more, even very responsive to private interests and to bottom-up participatory processes. But the actual means of implementation ultimately lie in the public sector. As mentioned above, the courts in various countries have pointed out the risk of "selling planning powers". There are also issues of legitimacy and democracy in moving powers and decision-making away from democratically elected governments and the machinery of the civil service.

Under this understanding of urban planning, participatory workshops with neighbourhood women, virtual collaborative platforms set up to discuss specific planning topics, and other means for citizen involvement in the process, are important tools and means to enable the participation of additional voices, but they do not qualify properly as urban planning on their own, separate from formal planning systems. In this respect, I would argue that some experiences sometimes considered to be gender mainstreaming which fall within this kind of activity cannot be properly understood as such. Because of their very limited scope, they fail to address the mechanisms of planning in any structural way, with only a minimal impact on routine practices and decision-making processes.

A second presupposition also addresses what is, or rather should be, gender mainstreaming within the specific field of spatial planning. Gender mainstreaming should also be a structural policy with effective means of implementation that addresses relevant topical areas within the particular sectoral policy in which it is being applied. This implies some actions that are sufficiently strong to have some transformational impact, combining a significant relevant number of measures in the topical areas relevant to planning (i.e. housing, transportation, facilities, public space, workspace, etc.). It should combine cross-cutting measures with measures specifically addressing women in the diversity of their urban needs.

The rather successful experience at mainstreaming gender in European science policy of the last two decades suggests that in order to be effective such policy requires, as basic inputs: sufficient gender expertise within the organisation, capacity building programmes, and specific units in charge of implementation with the capacity to both develop their own initiatives and to support the work of all other relevant departments (Abreu *et al.*, 2017). Political strategising by key actors mobilising support for the policy, and able to act effectively within policy processes, is also important to substitute for the required top-level support, when the latter does not exist.

To be effective, gender mainstreaming approaches in city and spatial planning should meet a number of requirements, amongst which the following can be highlighted:

- Measures, structures and processes for both positive action and cross-cutting action
- Institutional mechanisms to integrate the gender dimension within the internal procedures of institutions, throughout all processes and at all levels

- Expert knowledge in the matter, i.e. combining town planning and gender, including the creation and transmission of new expert knowledge and its integration within administrative structures and decision-making processes
- A broad range of measures of different types encompassing both small- and large-scale action, in a sufficient number of diversified areas of action, appropriate to the particular context and not limited to only one or two (such as security/safety in public areas or housing for victims of gender violence)
- A sufficient number of transforming measures involving structural changes, thus contributing to their institutionalisation and long-term change in organisations, so that integration of gender within public policies will be sustainable in the long term and not dependent upon the presence of specific individuals in decision-making positions in favour of this issue
- Realism in terms of what is feasible; not trying to implement an overly ambitious programme that is impossible to put in practice given the realities of the specific situation in terms of staff capacity, resources or political backing
- Legislative changes which ensure continuity over time and structural change
- Pilot projects as instruments for innovation and learning, as well as tools for awareness raising and visibility
- Intersectionality in all its dimensions, which is not to be confused with an interpretation of diversity that subsumes gender into a list of potential characteristics of discrimination that end up blurring the specificity of gender inequality

The new generation of equality policies in spatial development, which comes out of this new setting, must lastly build on solid foundations developing institutional capacity to impel the structural changes needed to achieve the ambitious goal set down in the introduction to the New Urban Agenda: to achieve gender equality and empower all women and girls, ensuring full and efficient participation and equality for women in all fields (United Nations, 2016).

Notes

1 There is a rich literature on gender in development which can bring interesting insights also to Northern countries. For example, Women's Major Group (2013) contains many practical illustrative examples from around the world, while UN Women (2014) is a guidance note providing very useful analysis and recommendations on process issues, including relevant illustrative examples in box format.

2 https://unfccc.int/resource/docs/2015/cop21/eng/l09r01.pdf.

3 See the argument in favour of universal access to the use of washing machines; the challenges that this entails, for the use of energy, climate change, and also for the construction of urban infrastructure; and the limits that exist for this with the current levels of energy consumption in other sectors, such as mobility in private vehicles (Sánchez de Madariaga, 2014).

4 For this specific trade-off between gender equality and environmental impacts in the transportation sector in the US case see Blumenberg (2016).
5 www.worldurbancampaign.org/general-assembly-partners.
6 The detailed non previously published information to the City of Vienna is contributed by Eva Kail.
7 See a detailed description of the projects run and their impact in Irschik and Kail (2013) and Damyanovic, Reinwald and Weikmann (2013).
8 This argument is further developed in Sánchez de Madariaga (2013).
9 The authors of this chapter were commissioned by the Basque Government to draft a document with policy proposals and practical measures addressing gender to be integrated as part of the DOT. These recommendations were fully integrated in the final DOT adopted in 2018 (Gobierno Vasco, 2018).
10 www.garraioak.ejgv.euskadi.eus/r41-430/es/contenidos/evento/urbanismo_inclusivo/es_urbincl/adjuntos/seminario_es.pdf.
11 www.eudel.eus/destacados/berdinsarea/. The guide *Ordenanza Marco en el ámbito de la igualdad entre mujeres y hombres en la CAPV* provides practical tools for municipal planning departments. www.eudel.eus/es/publicaciones/libros.
12 Planning mechanisms allowing for negotiated public-private partnerships are normally designed for complex, big projects. In France the Zones d'Aménagement Concerté (ZACs), in the US the American Development Agreements, and in the UK the planning gain, are used for this purpose. The legal limitations to using such mechanisms have been extensively studied by Renard (1988) and Gaudin (1996) for the French case; Johnston *et al.* (1990), Altshuler (1993), Delaney (1993) for the US, and Kayden (1988) and Smith (1988) for the UK.
13 The international development community uses the term "grassroots associations" or "grassroots women" to refer to people participating in such bottom-up processes.

References

Abreu, A., Faith, B., Hrynick, T., Sánchez de Madariaga, I., Spini, L. and Waldman, L. (2017). *Pathways to success. Bringing a gender lens to the scientific leadership of global challenges*. Paris: International Council for Science ICSU.

Alterman, R. (Ed.) (1988). *Private supply of public services. Evaluation of real estate exactions, linkage and alternative land policies*. New York/London: New York University Press.

Altshuler, A.A. (1993). *Regulation for revenue. The political economy of land use exactions*. Cambridge, MA: Brookings Institution.

Blumenberg, E. (2016). Why low-income women in the US still need automobiles. *Town Planning Review* (special issue of *Planning the Gendered City*, edited by I. Sánchez de Madariaga and M. Neuman), 87(5), 525–545.

Buckingham, S. and Le Masson, V. (Eds.) (2017). *Understanding climate change through gender relations*. London: Routledge.

Chapman, L. (2007). Transport and climate change. A review. *Journal of Transport Geography*, 15, 354–367.

Council of Europe (1998). *Gender mainstreaming. Conceptual framework, methodology and presentation of good practices*. Strasbourg: Council of Europe.

Damyanovic, D., Reinwald, F. and Weikmann, A. (2013). *Gender mainstreaming in urban planning and urban development*. Werkstattbericht, Nr. 130A. [Online] Vienna: Urban Development Vienna, Municipal Department 18 (MA 18) – Urban Development and Planning. Available at: www.wien.gv.at/stadtentwicklung/studien/pdf/b008358.pdf [Accessed 9 May 2018].

Davidoff, P. (1965). Advocacy and pluralism in planning. *Journal of the American Institute of Planners*, 31(4), 331–338.

Delaney, J. (1993). Development agreements. The road from prohibition to "Let's make a deal!". *The Urban Lawyer*, 25(1), 49–67.

Eudel – Asociación de Municipios Vascos (2012). *Ordenanza Marco en el ámbito de la igualdad entre mujeres y hombres en la CAPV (Model ordinance for equality between women and men in the Autonomous Community of the Basque Country)*. [Online] Eudel. Available at: www.eudel.eus/es/publicaciones/libros [Accessed 3 June 2018].

Gaudin, J.-P. (1996). La contractualisation des politiques urbaines (The privatisation of urban policies). In: P. Genestier, *Vers un nouvel urbanisme. Faire la ville, comment? Pour qui? (Towards a new urbanism. Building cities. How? For whom?)*. Paris: La documentation Française, 29–38.

Gobierno Vasco (2018). *Revisión de las Directrices de Ordenación Territorial de la Comunidad Autónoma del País Vasco DOT (Revision of the regional plan of the Autonomous Community of the Basque Country)*. [Online] Vitoria: Departamento de Medio Ambiente, Ordenación Territorial y Vivienda. Available at: www.euskadi.eus/modificacion-de-las-dot-en-lo-relativo-a-la-cuantificacion-residencial-2014/web01-a2lurral/es [Accessed 9 May 2018].

Houairou Commission and Cities Alliance (2016). *Engendering the New Urban Agenda. Comments on the Zero Draft*. New York City: Houairou Commission and Cities Alliance.

Irschik, E. and Kail, E. (2013). Vienna. Progress towards a fair shared city. In I. Sánchez de Madariaga and M. Roberts (Eds.), *Fair shared cities. The impact of gender planning in Europe*. Farnham: Ashgate, 193–230.

Jarvis, H., Cloke, J. and Kantor, P. (2009). *Cities and gender. Critical introductions to urbanism and the city*. London: Routledge.

Jáuregui, A. (2012). Informe sobre el grado de desarrollo de las políticas de igualdad de género en el planeamiento urbanístico municipal de la CAPV (Report on the degree of implementation of gender equality policies in municipal planning in the Basque Country). In: M. Apodaka Ostaikoetxea, *Urbanismo Inclusivo. Las calles tienen género (Inclusive planning. Streets are gendered)*. Vitoria: Gobierno Vasco, 75–88.

Johnston, R.A., Schwartz, S.I., Wandesforde-Smith, G.A. and Caplan, M. (1990). Selling zoning. Do density bonus incentives for moderate-cost housing work? *Land Use Law & Zoning Digest*, 42(8), 3–9.

Kayden, J. (1988). Planning gain. Developer provision of public benefits in Britain. In: R. Alterman (Ed.), *Private supply of public services. Evaluation of real estate exactions, linkage and alternative land policies*. New York/London: New York University Press, 163–172.

Kronsell, A. (2017). The contribution of feminist perspectives to climate governance. In: S. Buckingham and V. Le Masson (Eds.), *Understanding climate change through gender relations*. London: Routledge, 104–120.

Ley 4/2005, de 18 de febrero, para la Igualdad de Mujeres y Hombres del País Vasco (Law for Equality between Women and Men of the Basque Country).

MacGregor, S. (2017). Moving beyond impacts. More answers to the "gender and climate change" question. In: S. Buckingham and V. Le Masson (Eds.), *Understanding climate change through gender relations*. London: Routledge, 15–30.

Mazey, S. (2001). *Gender mainstreaming in the EU. Principles and practice*. London: London European Research Center, University of London.

Moser, C. (2006). *Asset-based approaches to poverty reduction in a globalized context*. Global Economy and Development Working Paper, 1. Washington, DC: The Brookings Institution.

Ranson, M., Morris, L. and Kats-Rubin, A. (2014). *Climate change and space heating energy demand. A review of the literature*. [Online] Washington, DC: Environmental Protection Agency NCEE. Available at: www.epa.gov/sites/production/files/2015-01/documents/climate_change_and_space_heating_energy_demand.pdf [Accessed 9 June 2018].

Reeves, D., Parfitt, B. and Archer, C. (2012). *Gender and urban planning. Issues and trends*. Nairobi: UN-Habitat.

Renard, V. (1988). Financing public facilities in France. In: R. Alterman (Ed.), *Private supply of public services. Evaluation of real estate exactions, linkage and alternative land policies*. New York/London: New York University Press, 173–181.

Sánchez de Madariaga, I. (2016). *Perspectiva de género en las Directrices de Ordenación del Territorio Vasco. Propuestas de Acción (Gender mainstreaming in the regional plan of the Basque Country. Proposals for action)*. Madrid: Cátedra UNESCO-UPM de Género, y Departamento de Medioambiente y Política Territorial del Gobierno Vasco.

Sánchez de Madariaga, I., (2014). *Access to washing machines. An indicator for measuring advance in the global agendas on gender, sustainability and urban quality*. [Online] UN-Habitat. Available at: http://unhabitat.org/washing-machines-indicator-for-measuring-advances/ (Accessed 3 June 2018).

Sánchez de Madariaga, I. (2013). *Mobility of care. Introducing new concepts in urban transport*. In I. Sánchez de Madariaga and M. Roberts (Eds.), *Fair shared cities. The impact of gender planning in Europe*. Farnham: Ashgate, 33–48.

Sánchez de Madariaga, I. and Boccia, T. (2016). Engendering Habitat III. Facing the global challenges in cities. *TRIA* (special issue *Engendering Habitat III. Facing the global challenges in cities*, edited by S. De Gregorio Hurtado and I. Novella Abril), 9(2), 13–22.

Sánchez de Madariaga, I. and Raudma, Tiia. (2011). *Structural change of research institutions. Enhancing excellence, gender equality and efficiency in research and innovation*. Brussels: European Commission.

Sánchez de Madariaga, I. and Roberts, M. (Eds.) (2013). *Fair shared cities. The impact of gender planning in Europe*. Farnham: Ashgate.

Smith, B. (1988). California development agreements and British planning agreements. The struggle of the public land use planner. *Town Planning Review*, 59(3), 277–287.

United Nations (2018). *Progress on the implementation of the New Urban Agenda Report of the Secretary-General*. New York: General Assembly.

United Nations (2016). *New Urban Agenda. Quito Declaration*. [Online] New York: UN General Assembly. Available at: www2.habitat3.org/file/537306/view/591158 [Accessed 9 May 2018].

United Nations (2015a). *Adoption of the Paris Agreement*. [Online] Paris: FCCC Conference of the Parties. Available at: https://unfccc.int/resource/docs/2015/cop21/eng/l09r01.pdf [Accessed 9 May 2018].

United Nations (2015b). *The Sendai Framework for Disaster Risk Reduction 2015–2030*. [Online] Geneva: UNISDR. Available at: www.unisdr.org/files/43291_sendaiframeworkfordrren.pdf [Accessed 9 May 2018].

United Nations (2015c). *Transforming our world. The 2030 Agenda for Sustainable Development*. [Online] New York: UN General Assembly. Available at: https://sustainabledevelopment.un.org/content/documents/21252030%20Agenda%20for%20Sustainable%20Development%20web.pdf [Accessed 9 May 2018].

UN Women (2014). *Gender mainstreaming in development planning*. New York: UN Women.

Women's Major Group (2013). *Gender equality, women's rights and women's priorities. Recommendations for the proposed Sustainable Development Goals (SDGs) and the Post-2015 Development Agenda*. New York: Women's Major Group.

8 Contemporary challenges in spatial development

Susan Buckingham and Anastasia-Sasa Lada

Introduction

As the foregoing chapters have demonstrated, spatial development is not renowned for its gender sensitivity. While the projects identified indicate how a sensitivity to gender difference, and attempts to include more women as professional and lay participants, can make for more liveable cities for everyone, these developments will be – indeed are being – tested by cross-border challenges to societies in Europe, as elsewhere. We consider three of these challenges here: the effects of and need to address climate change, forced migration of refugees, and austerity. These challenges are not mutually exclusive, indeed, arguably, they can be strongly interlinked (for example, a severely degraded environment caused by a changing climate forces international migrations either directly or, as a result of conflict, indirectly). These challenges are qualitatively and spatially different to those discussed elsewhere in this book in that they are international in scope. Climate change is a global phenomenon produced by cumulative actions in different places – particularly in rich countries with high levels of consumption. The effects are most acutely felt in places remote from where the contributing pollutants were first released. The most recent financial crisis, while experienced most acutely by poor communities within affected countries and by poorer countries, is attributable to banks and other financial institutions far from those who are expected to adjust their economies through reduced welfare, pensions, wages and/or public services. Forced migrations as a result of climate change, financial crises, war and other catastrophes frequently cross international borders although societies which are complicit in producing the circumstances provoking these migrations are often unwilling hosts. While the city or other urban areas remain the bases for spatial development, these development units are increasingly revealed to be inseparable from the areas which surround them, supply them and provide opportunities for exporting material of various use and value.

In this chapter we use two concepts which aid an understanding of how spatial development is gendered – the relationship between public and private spaces, and the commons – to try to understand the impacts of the three global challenges on gender. If spatial development has failed to address social – particularly gender – inequality through its policies, then those challenges with a more global reach will only exacerbate existing inequalities. Conversely, attending to structural equality and justice stimulates alternative development strategies which, at their best, can prevent problems emerging in the first place, and can provide alternative and fair solutions to problems when they do manifest.

Public and private spaces

Historically, feminist analysis focused on women's association with private-domestic spaces, and how their incursion into public space had been seen variously in terms of a "guest" in a normatively masculine world, potentially transgressive, threatening or inviting threat (Greed, 1994; McDowell, 1999). While this has changed positively in some places – notably the Nordic countries – there remain, to a greater or lesser degree, gendered differences in uses of and perceived rights to space across Europe. One feminist approach interprets a division of labour in which women are employed in the public sphere in times of productive growth, but are the first to be expelled from paid work when jobs are scarce. That women have borne the brunt of so-called austerity is true across Europe: not only have women been more likely to have lost paid work in the public sphere, but they have also been expected to take up caring work in their homes and communities when this has been withdrawn by the state. However, having access to their own private space is increasingly challenging for women. Recent research collected by FEANTSA[1] and the Fondation Abbé Pierre[2] has revealed that homelessness amongst women is increasing, as is homelessness amongst people with a migration background (FEANTSA and Fondation Abbé Pierre, 2018). Numerically, only Finland amongst EU countries recorded a decrease in homelessness between 2010 and 2017, with England and Ireland recording the highest increases (169 per cent and 140 per cent, respectively). This means that the number of women without a home is rising numerically as well as proportionally.

Recognising that the public/private divide is too crude a dichotomy, Horelli and colleagues conceptualised a "third space" of community and neighbourhood spatial activity which informed their Women and Planning Toolkit (Horelli, Booth and Gilroy, 2000). In this third, or intermediate, space, women were engaged in unpaid, and some paid, work (in, for example, faith-based institutions, schools, community centres, charities, as well as less formally). Mokkaram Abbas writes about the *mutanazahat*-semi-public/semi-private spaces (and space-times, as in women-only swimming pool sessions) outside the entirely public realm of the city, in which women can meet with neighbours, friends and extended family away from the male gaze (Abbas and Heur, 2014). While she writes about Nablus in Palestine, and also gives examples from elsewhere in the Middle East, the establishment of Muslim communities in European cities challenges spatial development to provide culturally appropriate safe spaces for women, without reinforcing spatial constrictions on women's and girls' lives. The Jagonari Centre[3] in East London, designed by Matrix, a women's collective architectural practice in London in consultation with local South Asian women, was one example of a European *mutanazahat* which has served and empowered South Asian women from different cultural and religious backgrounds. Sadly, the Jagonari Centre closed in 2015.

The commons

Neo-liberalism has also spawned the privatised public space of shopping malls, office and entertainment complexes, which exclude those who do not have the resources to use them, or right to work in them. This is yet another step in the process of "enclosure" which has systematically deprived people who depend on the "commons" for their basic needs, and which now sees life-sustaining resources, such as water and energy, controlled by a small number of multinational organisations.

One response to this has been the activity of "commoning", through which people are claiming back control over their neighbourhood and their future (Linebaugh, 2007). This varies in scale from reclaiming derelict land to create a community garden (for examples of women's community gardening projects see WEN, undated), to campaigning for common holding of land, resources and property. On a more prosaic note, "sharing economies" are being proposed to minimise the production of goods, and maximise their use value (for example, through car and bike-sharing, repair cafes, tool leasing and so on). More examples are given in Duncan McLaren and Julian Agyeman's book "Sharing cities", including Amsterdam's designation as "Europe's first Sharing City" (McLaren and Agyeman, 2015: 250). One example of claiming private land for common purpose has been through the Occupy movements which arose in cities most affected by the economic financial collapse from 2008 onwards. Squats and encampments of temporary stages, lecture halls, information booths, community kitchens and dormitories demonstrated the impact that a privatised public was having on disadvantaged communities.

One argument which has held considerable sway since its development in the 1960s was the "Tragedy of the Commons", in which Garrett Hardin (1968) claimed that the commons – as it was owned by no-one – was cared for by no-one, and therefore common users would maximise their short-term personal advantage, in competition with other users who would be expected to do likewise. The solution to properly manage the commons was seen to be "a coercive force", "private enterprise system … or socialism" (Hardin, 1968), "iron governments, perhaps military systems" (Heilbroner, 1974), "public agencies, governments or international authorities" (Ehrenfeld, 1972) or "public control" (Carruthers and Stoner, 1981). All these quotes are from Elinor Ostrom (1990: 9). Any of these corrective regimes would, without doubt, be deeply masculinist and regressive to gender – indeed any – equality. This has been evident through the response of international organisations to a perceived threat to commons such as forest. This has prompted organisations such as the World Bank to put land under private protection to the extent of excluding the humans whose livelihoods depended on its products, thereby creating refugees (Isla, 2009; Federici, 2011). However, Elinor Ostrom, the only woman to have won a Nobel prize in economics (which she shared with Oliver Williamson in 2009 for their work on the collective management of resources), proposes an alternative possibility of collective self-organisation and self-governing, which she supports with examples. Silvia Federici takes this further by arguing that we must produce ourselves "as a common subject". While she specifies that women's social reproductive work forms the basis of the commons, and commoning, she also demands that the artificial separation of productive and reproductive work needs to end. While this separation persists, there is a dissonance between a structural category of gender, finding an expression in what constitutes work, and a process category, which enables us to break down barriers between who does what work and how we value different forms of work. (See this volume's Chapter 1 for a discussion on process and structural categories of gender.)

Conceptualisations of the commons and commoning, including from a feminist perspective, can be helpful in understanding the challenges we are considering in this chapter, and in proposing some ways forward for future spatial development.

Below, we set out the context of each of the three challenges we have identified; we will explain how each is gendered, and we will explore how a gender-sensitive approach to planning for these challenges can make a difference. For many reasons, gender (and other social) justice/equality is integral to dealing with these challenges.

Climate change

The emission of greenhouse gases which is producing climate change is caused by production and consumption activities using the global commons as free waste sinks. These activities are generally the product of highly masculinised milieus, as are the strategies to manage these emissions, which are developed in gender-stereotyped professions and largely without regard to how activities, impacts and mitigation strategies are gendered. After presenting greenhouse gas (GHG) emissions and climate data for Europe, this section will review how climate change in Europe is gendered, and draws attention to how a gendered approach to climate change mitigation may be more successful than strategies developed by the same systematic processes which have led to the situation in which we find ourselves. It should be pointed out that the 2°C temperature reduction agreed by the Paris Agreement (UNFCCC, 2015) as the target for global warming is highly contentious, and not guaranteed to deliver us from catastrophic global warming. Joni Seager traces the 2°C desideratum to the economist William Nordhaus, who explored the trade-off point between economic growth and environmental considerations, which are both, as she points out, dealt with in highly gendered ways in which women are particularly disadvantaged (Seager, 2009). Climate change negotiations incorporated a gender dimension rather late, bearing in mind that the United Nations Conference on

Environment and Development in 1992 and the Beijing Women's Conference in 1995 had both explicitly linked the need for gender equality and environmental sustainability to be considered indivisibly. The women's lobby eventually prompted the UNFCCC Conference of the Parties (COP) to integrate a gender perspective in 2011, a delay which some consider "difficult to justify" (Morrow, 2017: 33). The Paris Agreement (UNFCCC, 2015) recognises gender equality (along with the equal rights of other disadvantaged groups), and parties to the Agreement require adaptation and capacity building to be gender-responsive, reflecting the lobbying work of many international and civil society organisations, including WEDO (Women, Environment and Development Organisation), WECAN (Women's Earth and Climate Action Network) and Genre en Action (Buckingham and Le Masson, 2017). Women remain a minority in state climate negotiating delegations and in the UNFCCC constituted bodies (Morrow, 2017).

Greenhouse gas emissions

The EU's target is to reduce GHG emissions by 20 per cent by 2020 and by 40 per cent by 2030 compared with 1990. Its own statistics claim that this has been surpassed with 2015 greenhouse gas emissions in member states down overall by 22 per cent compared with 1990 levels, representing an absolute reduction of 1,265 million tonnes of CO_2 equivalents. In general, the trend has been consistent, with the sharpest declines linked to the global financial and economic crisis and resulting reduced industrial activity. However, in 2015, GHG emissions increased for the first time since 2010 by 0.6 per cent (28.1 million tonnes of CO_2 equivalents) compared to 2014. This warns of the necessity to delink economic growth from energy production, or, indeed, of the problems of prioritising economic growth over other forms of prosperity (Jackson, 2009).

The largest source of GHG emissions is fuel combustion and fugitive emissions from fuels (excluding transport), although its share has decreased since 1990 (55 per cent of EU-28 greenhouse gas emissions in 2015). Fuel combustion for transport (including international aviation) has increased its contribution significantly since 1990 (23 per cent of EU-28 greenhouse gas emissions in 2015). GHG emissions from agriculture contribute 10 per cent to EU-28 total greenhouse gas emissions; industrial processes and product use contribute another 8 per cent; management of waste contributes 3 per cent. The EU-28's total carbon footprint was equal to an unsustainable 7.2 tonnes CO_2 per person in 2014 (European Commission, 2018).

European temperature

Largely as a result of these emissions worldwide, the average annual temperature over European land areas increased by 1.56–1.61°C in 2007–2016 relative to the pre-industrial period, making it the warmest decade on record. The warmest year in Europe since instrumental records began was 2014 with 2015 and 2016 close behind. Moreover, climate reconstructions show that summer temperatures in Europe in the last three decades (1987–2016) have been the warmest for at least 2,000 years, and that they lie significantly outside the range of natural variability. This has been particularly noticeable in the summer over the Iberian Peninsula, across central and north-eastern Europe and in mountainous regions. Warming since the 1960s has been strongest and most significant over Scandinavia, especially in winter. A pattern of greater frequency and duration of heat extremes across Europe has been detected and is projected to continue (European Environment Agency, 2017).

Gendered impacts

While women are not exclusively worst affected in climate-related/induced disasters, the World Health Organisation (WHO) refers to evidence which suggests that they are more vulnerable in heatwaves and flooding. For example, more women than men died during the 2003 European heatwave. It cites a majority of European studies which have shown that women are more at risk, in both relative and absolute terms, of dying in a heatwave. While there may be some physiological reasons for an increased risk among elderly women, social factors (such as living alone) can also be important in determining the risk of negative impacts of heatwaves (WHO, 2010).

Unpaid care work tends to increase in times of acute environmental stress. Tapsell and Tunstall (2001) as well as Tapsell, Tunstall and Wilson (2003) identified that women have the main responsibility for, and probably a greater emotional investment in, the home than men, and usually have the primary responsibility for the care of children and the elderly, as well as for getting the home back to normal after a flood, as Maureen Fordham's work has also shown in flooded areas of the UK (Fordham, 1998). Women may also be more ready to admit to feelings of stress, anxiety and depression and to seek medical help in the aftermath.

It has also been established that violence against women escalates in times of environmental stress. Margaret Alston's account of a catastrophic forest fire in Victoria, Australia, reports a majority of women interviewees in one study "experienced violence, many for the first time" (Alston, 2017: 140). This study, five years after the "Black Saturday" of the fires, found gendered responses to the bush fires at every stage: from whether to leave (women) or stay (men); community cohesion provided by women running communal kitchens; the frustration of men unable to have saved, and then to salvage their homes; the increase in marriage breakdown, alcohol and drug addiction; and domestic violence. Alston concludes with the call for gender mainstreaming to inform all stages of disaster relief and management, and to recognise women's and men's different responses to disaster. With heatwaves, bushfires and flooding increasingly affecting Europe, as elsewhere, spatial development urgently needs to understand gendered roles, approaches and attitudes in order to ensure not only that the environmental impacts do not fall disproportionately on women, but that reconstruction also responds to the various needs and experiences of women and men, as Alston argues from this Australian example (Alston, 2017).

Gendered causes

Of those decisions contributing to the largest producers of GHG emissions, many are located in spatial development – itself made up of several highly masculinised professions, as previous chapters have indicated. Planning decisions and policies on health and education, particularly from the late 1990s for example in the UK and other European countries, though not in Switzerland, have consolidated high schools, hospitals and universities as well as shopping and business parks in locations not always well served by public transport. While there has since been recognition of how out-of-town centres impact negatively on town centres in the UK, and the National Policy Planning Framework has devised two tests for proposed developments to promote the vitality of town centres (UK Ministry of Housing, Communities and Local Government, 2014), it has not eliminated out-of-town centres which generate additional road traffic and GHG emissions, and add to the care work of many women and some men.

In arguing for a mobility of care, Inés Sánchez de Madariaga (2013) presents European data from the Harmonised European Time Use Survey which shows that women spend more than double the amount of time as men in caring and household work, and that without a transport system which recognises how important it is to minimise travel both the quality of women's lives and the quality of the environment are compromised. Online shopping may be seen as one solution for those with time constraints, but it is one in which women's needs and environmental impacts need to be carefully thought through together. Wrigley and Lambiri (2014) have identified UK trends of non-food online sales, with books, videos, music, computer games and electricals, and fashion and department stores providing the bulk of online sales, though with food adding the most recent growth. As more online shops offer delivery of individually packaged goods to convenient urban locations, as well as to home addresses, this is likely to have environmental impacts, although Wrigley and Lambiri have not researched this aspect of online shopping. While there is not yet much research on the environmental impacts of online shopping, one US-based research project found an increase in the burden of travel time, delay, average speed and greenhouse gas emissions on the urban transportation network (Laghaei, Faghri and Li, 2015). An Australian women's group, "1 Million Women", which aims to build a "movement of strong, inspirational women and girls acting on climate change", also draws attention to the substantial increase in demand for cardboard as a result of online shopping (1 Million Women, 2016).

How gender-sensitive planning could reduce GHG emissions

Can a focus on gender equality contribute to reducing GHG emissions and reducing the impacts of climate change? There are many examples of urban projects that have already tried to combine gender and environmental justice. Women's housing projects have been developed in northern European countries such as Sweden and the Netherlands, where older women living alone or younger mothers of dependent children can live more cooperatively, sharing kitchens, childcare facilities or tool sheds, reducing isolation and increasing energy efficiency by pooling some material resources. Examples have already been provided of the Frauen-Werk-Stadt (FWS) in Vienna and a second FWS project developed in the 2000s, which focused on living in old age (Damyanovic, Reinwald and Weikmann, 2013: 90). Such a focus has a particular resonance for women – who outnumber men in older age – in light of the impacts of the heat extremes summarised above.

This initiative inspired Berlin in 2001, when it legislated to require a consideration of gender relations and how these interact with other social, demographic and cultural features in city planning and building (Senatsverwaltung für Stadtentwicklung, 2011). Such planning strengthens community and sense of place by increasing the mixed use of buildings, accessibility, mobility and a focus on the compact city. Gender-sensitive housing includes attention to social and intergenerational mix, provides robust and flexible housing to minimise the number of moves a household needs to make, and conserves resources, including through energy efficiency. Such measures can make a contribution to reducing greenhouse gases and the city's carbon footprint. The city of short distances likewise emphasises the local, thereby minimising environmental damage and expanding opportunities for neighbourhood interactions (Sánchez de Madariaga, 2013).

Berlin has a rich tradition of alternative housing options with environmental and social benefits: for example, one co-housing website (cohousing-berlin, undated) reports 1,000 eco-housing units, in which households combine private spaces with communal facilities to different degrees with social and environmental benefits. Lidewij Tummers (2017) has researched co-housing across Europe to understand the symbiosis between gender equality and low carbon impact. Recognising the power of the built environment to reinforce particular behaviours and lifestyles, she concludes that the social disruption needed to achieve gender equality "may be the key to resolving the challenges that climate change presents in a just manner" (Tummers, 2017: 235). Of course, as ecofeminists have long argued, this requires a rethinking of gender relations to ensure that gender equality is achieved by changing the dominance of industrial masculinity that sets current male norms of behaviour, including their relationship with non-human nature.

Small-scale projects, designed collaboratively with the communities they are intended for, are likely to serve both women and the environment well when they take into account the complex lives and multiple tasks of people involved in the design process. Examples of this include Col·lectiu Punt 6 (undated) in Barcelona, which was set up in response to national legislation to include a gender perspective in the design of urban spaces and facilities, and Urbanistas (undated), a women-led network that works with professional planning organisations in the

UK and Australia to encourage "women to draw on both personal values and professional know-how to lead and deliver projects … to [achieve] equality and social justice".

At a much larger scale, the international "C40 Cities" – a network of cities formed in an attempt to address climate change – has, under the leadership of Mayor Anne Hidalgo of Paris, developed a "Women4Climate initiative" (Hidalgo, 2018), which includes a global mentorship programme for women, influencing future leaders, stimulating innovations in climate and sustainability, and raising awareness through research on gender, cities and climate (C40 Cities, 2017). However, it remains to be seen how far this is able to influence the agendas of megacities, and whether the initiative will continue once the C40 leadership rotates to a less feminist mayor (Hidalgo, 2018).

In summary, climate change is a result of the abuse of the commons by industrial-scale processes characteristic of a hyper-masculinised society. The impacts of climate change are felt in places and by people most likely to be furthest removed from the original polluters, who are so far able to remove themselves from the worst impacts. Spatial development needs strategies to adapt to a changed climate, just as it must plan for less polluting activities. In Europe this primarily relates to reducing dependency on fossil fuels and road transport, minimising the distance travelled between life's different activities, reducing consumption and the waste resulting from this, and creating energy efficiency in new buildings and established stock. Such development needs to recognise how these strategies impact on gender relations and strive for greater gender equality. This, of course, needs underlying policies to diminish the division between productive and reproductive labour, and their relative value, as suggested by Federici (2008) and as discussed above. The recognition of intermediate spaces, and the relation between this, public and private space, following Horelli, Booth and Gilroy (2000), also needs to be greater. The spatial correlate of this will be to develop spaces in which productive and reproductive activities can be shared and valued equally, and this can be expected to reduce carbon emissions. Some examples of small-scale women-led strategies suggest the potential for such spatial development, and collaborative exercises that stress the potential for a different kind of social practice which, as referred to earlier, Elinor Ostrom found prevalent in her research on collaborative economics. Without collaboration, the pursuit of competitive advantage leads to unforeseen global consequences, two more of which are considered next.

Forced migrations

Climate change is one reason which causes people from severely affected countries to seek asylum and refuge elsewhere – by some estimates displacing up to 1 billion people, but more conventionally 200 million people by 2050 (Christian Aid, 2007; IOM, 2008a, 2008b). This can be through food shortages and chronic environmental degradation, an increase in extreme weather events which generate mass displacement, or sea-level rise destroying low-lying coastal areas that are home to millions of people who will have to permanently relocate. While the human, and specifically gendered, consequences of these displacements are poorly studied, if at all, it can be expected from other – non-climate induced – disasters that women and children are particularly vulnerable to "sexual violence, sexual exploitation and abuse, trafficking, and domestic violence in disasters" (International Federation of Red Cross and Red Crescent Societies, 2007: unpaged). Adolescent girls report especially high levels of sexual harassment and abuse in the aftermath of disasters and complain of the lack of privacy in emergency shelters.

Other causes of forced migration include unequal economic conditions, which can effectively force people to move from a poor to a richer area in search of work and other resources, and political unrest – frequently the result of Western intervention in poor countries. This section reviews recent refugee and asylum-seeking migration across Europe before addressing further how migrant experiences are gendered, and how using a gender-sensitive framework could better manage migration and the impacts of migration in Europe.

Migration of asylum seekers into Europe

A European Parliament report on the reception of female asylum seekers and refugees notes that most asylum seekers arrive in Europe via Greece or Italy, although this is not where most refugees and asylum seekers register. According to the United Nations High Commissioner for Refugees (UNHCR), 1,015,078 people arrived by sea (mainly by the Mediterranean Sea) in the EU in 2015. Of these people 50 per cent were men, 19 per cent were women and 31 per cent were children, with a similar distribution in the first five months of 2016. As of March 2016, women, children and unaccompanied minors account for more than half of all refugees arriving in Greece, Italy, Malta, Spain and Cyprus altogether (European Parliament, 2016).

The European Parliament report also acknowledges that despite legal recognition of persecution based on gender, "women have, in general, more difficulties in proving their claim for asylum" (European Parliament, 2016: 13), particularly where the violence (domestic violence, female genital mutilation (FGM) and/or rape) they have experienced has no corroborating evidence.[4]

> This is also the reason why female applicants for asylum are less successful than male applicants. Their stories are considered to be less credible than those of men (who can provide actual proof for their persecution, for example membership of a certain party). The way in which women present their story is therefore even more important than for men.
>
> (European Parliament, 2016: 13)

However, women are likely to find telling their story more difficult than men, particularly where this includes sexual violence, and the situation is worsened when they have to recount their story to a man, or where they are accompanied by their children for whom there are no childcare facilities in reception centres (European Parliament, 2016).

Gendered implications for asylum seekers

The situation for women refugees and asylum seekers requires that their emergency housing is organised in a gender-sensitive way, including to prevent violence against women. For example,

> Overcrowded reception centres pose serious risks of (sexual) violence against women and children. In order to further improve the safety and privacy of women, private bathing and sanitation facilities should also be provided in reception centres, in particular in cases where the centres are overcrowded. Men and women's sanitary facilities should be strictly separated, as well as easy and safe to reach.
>
> (European Parliament, 2016: 14)

In addition, proper healthcare must be present at the reception centres, in particular for pregnant women, and women should also be informed or receive counselling about their reproductive health.

If asylum is granted, attention to gender sensitivity in longer-term housing and support is also necessary. The EU records concern that for some women, selling sex is seen as the only way in which they can access money or other resources to survive (European Parliament, 2016).

Gendered impacts of flows of asylum seekers into Europe

Where, as is the case in Europe presently, the number and proportion of men seeking asylum outnumbers women, there is likely to be an unevenly gendered impact on the societies to which asylum seekers migrate. The EU has warned of the unwiseness of not allowing the families of refugee men to join them (European Parliament, 2016). The case of large-scale attacks by mainly asylum-seeking men on women in Cologne in December 2015 has been used to illustrate this, although, as the *Economist* points out, large-scale events such as the annual Oktoberfest in Munich also routinely feature an upsurge in gendered attacks on women, which the police fail to deal with adequately (*Economist*, 2016).

Without becoming competitive about the relative rights of women and ethnic minorities, it is clear that some sensitivity is needed in making provision for the human right to a family life and the right to be safe, and appropriate spatial development is needed to accommodate refugees and asylum seekers in humane ways which also respect local communities. An initiative has been taken by local communities in the UK and Ireland which have joined a network of cities in which facilities for refugees can be provided, "building a culture of hospitality for people seeking sanctuary in the UK" (City of Sanctuary, undated: unpaged). More informally, Vaiou and Lafazani (2015) hold up the example of the Athens neighbourhood of Kypseli as a place where migrants have integrated well with long-term residents in a mutually supportive way.

Jonathan Darling argues that "it is important that resettlement and support services are appropriately funded, and situated within strong networks of communication between municipalities and other agencies". He quotes the Eurocities network: "no resettlement scheme can be successful if a reception place is not available at local level, if local civil society organisations are not involved and if awareness-raising and consensus building amongst the local population are not properly managed" (Darling, 2015: unpaged).

As an illustration of how solutions to contemporary challenges can be interlinked, some of the most innovative provisions for refugee housing build in elements of environmental sustainability and work training. In Munich, for example, a local development alliance of creatives, activists and organisations working with refugees, created a social co-operative to renovate three inner city buildings for occupation by refugees. One of the architectural aims of the project is to repair as much of the structure and decorations as possible, including window frames, tiles, pipes, wooden floors. Anyone involved in the building project – volunteers, refugees – has their participation recognised as a professional qualification in the form of a certified apprenticeship. Provision is made for accommodation for between 40 and 45 residents, including for families and single-parent (mother) families, in two of the buildings. The third is reserved for shared space (a communal kitchen, legal and medical advice office, education facilities). Provision is also made for a merging of inside and outside space which creates a kind of intermediate – or *mutanazahat* – space referred to earlier. In Bremen, temporary accommodation for refugees has been created from prefabricated containers arranged as two-bedroom apartments in three courtyard structures. Providing housing for 162 residents in seven different buildings, the development is close to education facilities, and also includes play spaces, an orchard, bicycle storage and outdoor seating areas. The courtyards, providing a transition between private and public spaces, and emulating a village structure, are a deliberate attempt to recreate traditional residential forms of Arab villages (Cachola Schmal, Scheuermann and Elser, 2017). These are both clear examples of spatial development responses which address a number of contemporary challenges simultaneously, while sensitive to the needs of women and men.

The structures and arrangements referred to above are newly built, but older neighbourhoods can also provide for communities in which migrants and older established communities can, over time, integrate. Dina Vaiou writes in the *Athens Social Atlas* how a form of benign neglect on the part of Athens planners has created a neighbourhood space in Kypseli in which

> support between local and migrant men and, most importantly, women is mutual and acquires a major significance at times of crisis and of dramatic cuts in salaries and services, since it redefines in various ways the meaning of neighbouring and neighbourhood. Here the social life of migrant women, the sense of companionship and acceptance by local women as well as their

> everyday practices contribute to a sense of belonging in the neighbourhood and the city.
>
> (Vaiou and Lafazani, 2015: unpaged)

The next section will examine the situation in Greece in more detail, particularly from the vantage point of how it has been impacted by austerity.

Austerity

The story of the 2008 economic crisis is now well known. The bursting of the housing bubble in the US in 2007 and the fall of Lehman Brothers in 2008 caused the near collapse of the global financial system. This triggered the deepest and most protracted global economic crisis since 1929 and the second major crisis that the US has exported to the world.

Although the interconnectedness of financial markets diffused the crisis rapidly from the US to other parts of the world, this occurred on different time scales, with different intensities and followed different patterns, according to regional and national specificities in both transmission mechanisms and in their pre-existing economic imbalances. For instance, in Europe, the global crisis has seriously destabilised the Eurozone with a series of sovereign debt crises that erupted in its periphery, mainly in the so-called south (Greece, Spain, Portugal and Ireland).

Austerity has become the new principle for public policy in Europe and the US, as the financial crisis of 2008 has been converted into a public debt crisis. However, current austerity measures risk losing past progress towards gender equality by undermining important employment and social welfare protections and putting gender equality policy onto the back burner (Karamessini, 2014). Some feminist scholars such as Walby (2009) have argued that the principles underlying the aims and goals of institutions of financial governance are gendered; and gender inequalities in the governance of the financial architecture are part of the cause of the crisis. The common thread here is that the causes of the financial crisis are to be looked for in the wider power relations of society and not simply in narrow issues such as forms of financial regulation.

After presenting the gendered impacts of austerity in Europe, this section will reflect on the possibility of developing a gender-sensitive approach to spatial development and planning within the neo-liberal era and the new austerity public policies. Finally, Greece will be used as a case study, since Greece has been affected by austerity policies in many different ways, one of which is equity and gender relations.

Gendered impacts of austerity – gender in (the) crisis or what happens to women in hard times

The relationships between gender and austerity must be considered as historically and societally specific. Gender relations have evolved over time but following country-specific paths, and the impact of the austerity programmes will depend both on the form and the robustness of gender relations in a specific country context and on the particular form that the austerity programmes take in interaction with wider labour market, social and political institutional arrangements.

The gender impact of austerity is also likely to reflect differences between the positions of men and women in relation to the family economy and to the welfare system. These differences include not only material and work relationships, such as contribution to family income, to caring work and access to benefits and public services, but also social norms with respect to the appropriate and expected gender roles. Women's presumed distinctive characteristics as carers first and labour force participants second have underpinned the notion that women may act as either a voluntary or involuntary labour reserve (Bruegel, 1979; Rubery and Tarling, 1982).

The full impact of austerity policies across Europe is still being assessed, but it is clear that their gendered impacts have been consistently severe, structural and manifold. They have also been, until now, under-researched and under-estimated.

In October 2012, a report by the European Women's Lobby (EWL, 2012) titled "The price of austerity – The impact on women's rights and gender equality in Europe" considered evidence of the impact of austerity measures on women's rights and gender equality in Europe. Based on data from EWL member organisations in 13 countries and recent research from a range of sources, it reveals that austerity policies in Europe undermine women's rights, perpetuate existing gender inequalities and create new ones, and hamper the prospects of sustainable and equal economic progress in Europe. The recent cuts in public spending in European Union (EU) member states, sanctioned by the European Commission, have the most impact on those who have little voice in economic decision-making: women, children and the elderly. Since the report, several European bodies and a number of researchers across Europe have focused on the gendered impacts of austerity (Manganara, 2014).

A project titled "Everyday Family Life in Austerity, 2012–2016, in Greater Manchester" by Sarah Marie Hall explored the impact of austerity on everyday family life, with a focus on gender, class and intergenerationality. This project shows that women bear the brunt of economic crises and that recession, austerity and economic changes imposed from above are not evenly felt or distributed across society. The retreat of the welfare state is creating gaps that disproportionally affect women in the care work necessary for social reproduction. Hall discusses the goals of an economic system that should fairly distribute caring responsibilities, labour and costs between women and men, and between families and the wider communities. She concludes that social inequalities are the symptom not the cause of everyday austerity (Hall and Ince, 2016).

Spatial development and planning in the neo-liberal era – reflecting on the Greek case

With the backdrop of the continuing global recession, the Greek economy seems to be no less than what was very early referred to as a convenient neo-liberal experiment. Harsh austerity measures, among them the dismantling of public services and social infrastructure and the shrinkage of public employment, have had severe implications in the real life of millions of citizens, while the persistent erosion of the urban planning system, along with the urgent sale of public property, have had almost no results in terms of the desired capital investment and land development. But, although the austerity measures did not deliver the anticipated economic results, they did deliver a new social and political situation in Greece. They produced a class of "nouveau poor", and turned a public of "Indignados" into a public of "Desperados" (Kaika, 2012).

Consequences permeate Greek society. In 2015, 39.9 per cent of the population in Greece was materially deprived and 35.7 per cent was at risk of poverty or social exclusion (the highest in Europe after Bulgaria), while 41.4 per cent of the households lived in poor housing conditions. The unemployment rate rose to 24.9 per cent, while the percentage soared above 38 per cent for youth. Overall, of the country's 11 million people, more than 3.5 million live at the edge of poverty.

The signs of the multifaceted crisis that Greece has gone through are most visible in urban areas and particularly in the two metropolitan regions of Athens and Thessaloniki. Following the collapse of the small and medium-sized economy, Greek cities face unforeseen decay of their physical and socio-economic tissue in commercial as well as in residential areas. The formerly internationally admired qualities of urban places with their small-grain mix of uses and subsequent 24-hour vitality and safety throughout the central and peripheral areas seem to be vanishing. At the same time, the number of unsettled refugees together with deprived and homeless people and beggars is rising in public space. It is in cities that the effects of economic adjustment programmes are more intensely felt. "Cities are therefore where austerity bites. However, never equally", as Peck (2012: 629) argues, drawing from the US experiences of austerity urbanism. In the European context, most prominently missing from the debate of the deepening of crisis is often the fact that the effects of austerity are unevenly distributed and refashioned on inequalities that preceded the current conditions in terms of gender, age, ethnicity, security of employment status, spatial differentiation and very often all these combined (Hadjimichalis, 2011; Karamessini, 2014; Svarna, 2014; Vaiou, 2014b).

It is important to point out that urban design processes in Greece are inscribed in a central planning system which is embedded in specific articles of the Greek Constitution, where the State is responsible for settlement development. Since the 1980s it has been constituted in practice through a hierarchical and centrally (i.e. governmentally) controlled succession of long-term plans from the scale of spatial planning to that of urban design. The development process in Greek cities was "worlded" by a political tradition in clientelism related to fragmented property and small capital investments, thus producing incremental and piecemeal growth of urban tissues of random architectural qualities. The resulting compactness, functional mix, cultural and social hybridity provided more-or-less for distinct safety and vitality as well as restricted segregation.

While large-scale urban design projects (mostly Olympic Games infrastructure) took place in exceptional statutory frameworks from the beginning of the 2000s, it is since 2010 that the whole of the urban planning and development process has come under neo-liberalisation scrutiny and dismantling attempts triggered by the so-called "Memorandum" regime (the framework of Economic Adjustment Programmes for Greece). In order to facilitate large-scale investment in regional and urban development and "fast-track" exploitation of precious public land by international corporate capital, new legislation enacted in July 2014 coined Special Spatial Plans, which were prioritised over former official plans. These transcend local democratic decision procedures considered to be obstacles to development and enhance neither social equity matters nor environmental sustainability issues. Furthermore, there has as yet been no detailed urban design guidance ordinance to ensure – or at least frame – qualities of urban form, place and living (Christodoulou and Lada, 2017).

Gender equality policy in times of austerity

It is ironic that the sovereign debt crisis and the definitive turn to austerity in 2010 coincided with the launch of the ambitious National Programme for Substantive Gender Equality 2010–2013 (General Secretariat for Gender Equality, 2010). This included all the prerequisites for the implementation of a fully fledged and all-embracing gender mainstreaming strategy at all levels of government (central, regional and municipal) for the first time in Greece, entirely co-financed by the European Social Fund. It also provided for the creation of a wide network of counselling centres for women and shelters for victims of domestic violence over the whole national territory. The Programme also encompassed measures to facilitate women's participation in employment and advancement at work (vouchers to employed/unemployed mothers for access to childcare services free of charge and to unemployed women for home-help services free of charge to the elderly and non-self-sufficient dependents), to encourage female entrepreneurship (positive action schemes) and promote the social integration of vulnerable groups of women (victims of trafficking and domestic violence, heads of single-parent households, women working under precarious forms of employment, etc.). It has, however, proved totally inadequate to compensate for the massive destruction of jobs, the unrestrained rise of unemployment, the loss of income and spread of poverty, and their disruptive effects on women's lives and those of their families, caused by "one of the biggest fiscal consolidation that any EU country has done over the past 30 years" (UNICRI, 2014).

Economic crises are especially hard on women (Seguino, 2009; Walby, 2009; Bettio *et al.*, 2013; Manganara, 2014; UNICRI, 2014) and the Greek crisis is no exception (Athanasiou, 2011; Avdela, 2011; Karamessini, 2014; Lyberaki and Tinios, undated; Vaiou, 2014a, 2014b). The effects of lay-offs, wage and pension reductions and the collapse of social services are felt by everyone, but women have less to lose and they are losing it faster. Those under 25 and over 65, single mothers and immigrants are especially vulnerable. With or without paid jobs, women are expected to care for small children who have to stay home due to the closing down of state nurseries, look after older and sick family members who no longer have access to social services, and exercise their homemaking skills so as to make the best of household resources. Having work while being married to a jobless husband often produces a sense of guilt, which women try to manage by intensifying their commitment to housework and family care. Violence against women at home is on the rise (Davaki, 2013). According to a recent study by the Department of Sociology of the Panteion University of Athens, the National Centre for Social Research and the Office of the State Ombudsman, prostitution has increased dramatically since the onset of the crisis and this is largely due to the fact that more and more women turn to sex work in order to support themselves and their children (BHMAgazino, 2015). However, the gendered aspects of the "Greek crisis" draw little public attention and this can only exacerbate work and income inequalities between women and men (Federici, 2008).

Opposed to this condition of un-negotiated facilitation of capital investment in the ongoing socio-economic and political crisis stand a series of informal resistances "from below" (Arampatzi, 2016; Vaiou and Kalantides, 2016). Against the almost total impairment of public investment in the urban environment, grassroots movements and solidarity initiatives are collectively reclaiming public urban places and thus reasserting their "right to the city". Despite their differentiated political standpoint – not always clearly supportive of urban entrepreneurialism and individualism or based on emancipation, democracy and equality (Athanassiou, Kapsali and Karagianni, 2015) – these citizen-led initiatives constitute emergent forms of contestation. An intrinsic aspect of these is the gender allocation among the volunteers and the dynamic, if unseen, engagement of women, which will hopefully ensure the endurance of many such ventures. The temporary and the spontaneous aspects of urban life and public space are gradually finding their place anew in the Greek cities, while their perspective in future official urban design processes remains obscure.

Concluding reflections

In this chapter we have discussed three processes: climate change, migration and austerity. They are global in provenance, scope and reach, and all impinge on the work of spatial development both separately and as they interact with each other. These processes provide the grounds for a critical re-engagement with, and a critical re-imagining of, who counts as part of the public; how the political is performed; how and where it takes place; what qualifies as political subjectivity, and how it is gendered, racialised and classed. Although they are usually marked under a "state of emergency" their discourses prove to be not exceptional but rather ordinary, systematic, canonical and foundational, producing dispensable and disposable populations, and, at the same time, producing and demarcating the normative codes of the human by regulating the (economic) vitality, affectivity, potentiality, embodiment, vulnerability and liveability of subjects.

Spatial development is not, on the whole, well attuned to consider gender relations and its inequalities and inequities, for example through gender-impact assessments or gender budgeting (though see Vienna – in this volume – and Andalucía – Junta de Andalucía (2018) – for path-breaking "gender+" initiatives). This lacuna inhibits planning bodies' abilities to deal with these challenges in gender-sensitive ways, thereby likely exacerbating gender inequality. We have used some examples to illustrate this gender blindness in order to emphasise how important sensitivity to gender (and other social inequalities) is for spatial development. We have also indicated some – albeit smaller-scaled – examples which demonstrate how gender considerations can be made.

As the example of the gender- and culturally sensitive *mutanazahat* has identified, gender-aware strategies need to finely balance responses to the existing realities of most women's lives (as prime carers of dependent family members, as unpaid housekeepers and as potentially vulnerable to male violence and abuse), while empowering women to live other kinds of lives in which care and domestic work, as well as professional work, is shared, and violence is minimised. Planners also need to be aware of the masculinist constructions of concepts such as *the commons* which have privileged competitive interpretations of use over more collaborative ones. Gender-sensitive spatial planning for global challenges, then, needs to question prevailing orthodoxies while introducing robust and embedded ways of assessing how gender equality is impacted by all policies, strategies and initiatives being considered to address these challenges.

Notes

1 FEANTSA: European Federation of National Organisations Working with the Homeless.
2 The Foundation Abbé Pierre supports better housing for the deprived.
3 The Jagonari Women's Resource Centre was designed by women for and in consultation with South Asian women in London's East End. The Matrix women's architects collective was approached by the Jagonari group of women in 1982 to design the space. The Jagonari Women's Education Resource Centre Ltd was eventually liquidated in 2015 having experienced financial difficulties, including the requirement to pay a market rent in a now fast gentrifying part of London. The building still exists, but is used for commercial childcare provision. For more information on the development of the space, see https://surveyoflondon.org/map/feature/453/detail/.
4 According to written evidence to the British Parliament, it is reported that 90 per cent of women seeking asylum are refused and that a greater proportion of appeals from women, than men, are unsuccessful (UK Parliament Home Affairs Committee, 2013).

References

1 Million Women (2016). *What's the environmental impact of your online shopping habits?* [Online]. Available at: www.1millionwomen.com.au/blog/whats-environmental-impact-your-online-shopping-habits/ [Accessed 20 July 2017].

Abbas, M. and Heur van, B. (2014). Thinking Arab women's spatiality. The case of "mutanazahat" in Nablus, Palestine. *Gender, Place and Culture*, 21(10), 1214–1229.

Alston, M. (2017). Gendered outcomes in post-disaster sites. Public policy and resource distribution. In: M. Griffin Cohen (Ed.), *Climate change and gender in rich countries. Work, public policy and action*. London: Routledge, 119–130.

Arampatzi, A. (2016). The spatiality of counter-austerity politics in Athens, Greece. Emergent "urban solidarity spaces". *Urban Studies*, 54(9), 2155–2171.

Athanasiou, A. (2011). Becoming precarious through regimes of gender, capital and nation. *Cultural Anthropology*, [Online] 28 October 2011. Available at: www.culanth.org/fieldsights/250-becoming-precarious-through-regimes-of-gender-capital-and-nation [Accessed 3 Mar. 2018].

Athanassiou, E., Kapsali, M. and Karagianni, M. (2015). Citizens' participation in urban governance in crisis-stricken Thessaloniki (Greece). Post-political urban project or emancipatory urban experiments? *RC21 International Conference on "The Ideal City: between myth and reality. Representations, policies, contradictions and challenges for tomorrow's urban life*. Urbino, 27–29 August 2015. [Online]. Available at: www.rc21.org/en/wp-content/uploads/2014/12/C2-Athanassiou-Kapsali-Karagianni.pdf [Accessed 3 Mar. 2018].

Avdela, E. (2011). Gender in (the) crisis or what happens to women in hard times. *Synchrona Themata*, 115, 8–17 (in Greek).

Bettio, F., Corsi, M., D'Ippoliti, C., Lyberaki, A., Samek Lodovici, M. and Verashchagina, A. (2013). *The impact of the economic crisis on the situation of women and men and on gender equality policies*. [Online]. Available at: www.researchgate.net/publication/303566528_The_Impact_of_the_Economic_Crisis_on_the_Situation_of_Women_and_Men_and_on_Gender_Equality_Policies [Accessed 3 Mar. 2018].

BHMAgazino (2015). *Prostitution in Athens of 2014*. [Online] 8 June 2015. Available at: www.tovima.gr/vimagazino/views/article/?aid=618417 (in Greek). [Accessed 3 Mar. 2018].

Bruegel, I. (1979). Women as a reserve army of labour. A note on recent British experience. *Feminist Review*, 3, 12–23.

Buckingham, S. and Le Masson, V. (2017). Introduction. In: S. Buckingham and V. Le Masson (Eds.), *Understanding climate change through gender relations*. London: Routledge, 1–12.

C40 Cities (2017). *C40 events. Women4Climate*. [Online]. Available at: www.c40.org/events/women4climate [Accessed 3 Mar. 2018].

Cachola Schmal, P., Scheuermann, A. and Elser, O. (2017). *Making Heimat. Germany, arrival country. Atlas of refugee housing*. Berlin: Hatje Cantz.

Christian Aid (2007). *Human tide. The real migration crisis*. [Online]. Available at: www.christianaid.org.uk/sites/default/files/2017-08/human-tide-the-real-migration-crisis-may-2007.pdf [Accessed 3 Mar. 2018].

Christodoulou, C. and Lada, S. (2017). Urban design in the neoliberal era. Reflecting on the Greek case. *Journal of Urban Design*, 22(2), 144–146.

City of Sanctuary (undated). [Online]. Available at: https://cityofsanctuary.org/ [Accessed 3 Mar. 2018].

cohousing-berlin (undated). [Online]. Available at: www.cohousing-berlin.de [Accessed 31 Oct. 2018].

Col·lectiu Punt 6 (undated). [Online]. Available at: www.punt6.org/en/ [Accessed 17 July 2017].

Damyanovic, D., Reinwald, F. and Weikmann, A. (2013). *Gender mainstreaming in urban planning and urban development*. Werkstattbericht, Nr. 130A. [Online] Vienna: Urban Development Vienna, Municipal Department 18 (MA 18) – Urban Development and Planning. Available at: www.wien.gv.at/stadtentwicklung/studien/pdf/b008358.pdf [Accessed 9 May 2018].

Darling, J. (2015). *Cities of sanctuary*. [Online]. Available at: www.opendemocracy.net/can-europe-make-it/jonathan-darling/cities-of-sanctuary [Accessed 3 Mar. 2018].

Davaki, K. (2013). *The policy on gender equality in Greece. Note*. Directorate-General for internal Policies, Policy Department C, Citizens' Rights and Constitutional Affairs (Ed.). [Online] Brussels: European Parliament. Available at: www.europarl.europa.eu/RegData/etudes/note/join/2013/493028/IPOL-FEMM_NT%282013%29493028_EN.pdf [Accessed 3 Mar. 2018].

Economist (2016). *Migrant men and European women*. [Online] 16 January 2016. Available at: www.economist.com/news/leaders/21688397-absorb-newcomers-peacefully-europe-must-insist-they-respect-values-such-tolerance-and [Accessed 3 Mar. 2018].

European Commission (2018). *Greenhouse gas emission statistics – carbon footprints*. [Online]. Available at: http://ec.europa.eu/eurostat/statistics-explained/index.php/Carbon_dioxide_emissions_from_final_use_of_products [Accessed 3 Mar. 2018].

European Environment Agency (2017). *Global and European temperature*. [Online]. Available at: www.eea.europa.eu/data-and-maps/indicators/global-and-european-temperature-4/assessment [Accessed 3 Mar. 2018].

European Parliament (2016). *Reception of female refugees and asylum seekers in the EU Case study Belgium. In-depth analysis*. [Online]. Available at: www.europarl.europa.eu/RegData/etudes/IDAN/2016/556946/IPOL_IDA(2016)556946_EN.pdf [Accessed 3 Mar. 2018].

EWL – European Women's Lobby (2012). *The price of austerity – the impact on women's rights and gender equality in Europe*. [Online]. Available at: www.womenlobby.org/ [Accessed 17 Nov. 2017].

FEANTSA and Fondation Abbé Pierre (2018). *Third overview of housing in Europe, 2018*. [Online] Brussels: FEANTSA. Available at: www.feantsa.org/download/full-report-en1029873431323901915.pdf [Accessed 12 July 2018].

Federici, S. (2011). *Feminism and the politics of the commons*. [Online]. Available at: www.commoner.org.uk/wp-content/uploads/2011/01/federici-feminism-and-the-politics-of-commons.pdf [Accessed 3 Mar. 2018].

Federici, S. (2008). *Precarious labor. A feminist viewpoint*. [Online]. Available at: https://inthemiddleofthewhirlwind.wordpress.com/precarious-labor-a-feminist-viewpoint/ [Accessed 3 Mar. 2018].

Fordham, M. (1998). Making women visible in disasters. Problematizing the private domain. *Disasters*, 22(2), 126–143.

General Secretariat for Gender Equality (2010). *National programme for substantive gender equality 2010–2013*.

Greed, C. (1994). *Women and planning. Creating gendered realities*. London: Routledge.

Hadjimichalis, C. (2011). Uneven geographical development and socio-spatial justice and solidarity. European regions after the 2009 financial crisis. *European Urban and Regional Studies*, 18(2), 254–274.

Hall, S.M. and Ince, A. (Eds.) (2016). *Sharing economies in times of crisis. Practices, politics and possibilities*. Routledge Frontiers of Political Economy. London: Routledge.

Hardin, G. (1968). Tragedy of the commons. *Science*, 162(3859), 1243–1248.

Hidalgo, A. (2018). Foreword. In: A. Hidalgo, *Why women will save the planet*. London: Zed Books/Friends of the Earth, xv–xvii.

Horelli, L., Booth, C. and Gilroy, R. (2000). *The EuroFEM toolkit for mobilizing women into local and regional development*. Helsinki: Helsinki University for Technology.

International Federation of Red Cross and Red Crescent Societies (2007). *World Disasters Report*. [Online] Geneva: International Federation of Red Cross and Red Crescent Societies. Available at: www.ifrc.org/PageFiles/99876/2007/WDR2007-English.pdf [Accessed 3 Mar. 2018].

IOM – International Organisation for Migration (2008a). *Climate change and migration. Improving methodologies to estimate flows*. IOM Migration Research Series, 33. [Online] Geneva: IOM. Available at: http://publications.iom.int/system/files/pdf/mrs-33.pdf [Accessed 3 Mar. 2018].

IOM – International Organisation for Migration (2008b). *Migration and climate change*. IOM Migration Research Series, 31. [Online] Geneva: IOM. Available at: http://publications.iom.int/system/files/pdf/mrs-31_en.pdf [Accessed 3 Mar. 2018].

Isla, A. (2009). Who pays for the Kyoto Protocol? Selling oxygen and selling sex in Costa Rica. In: A. Salleh (Ed.), *Eco-sufficiency & global justice. Women write political ecology*. London: Pluto Press.

Jackson, T. (2009). *Prosperity without growth. Economics for a finite planet*. London: Routledge (Earthscan).

Junta de Andalucía (2018). *Informe de Evaluación de Impacto de Género del Presupuesto para 2018 (Report for the Gender Impact Assessment of the Budget 2018)*. [Online]. Available at: www.juntadeandalucia.es/haciendayadministracionpublica/planif_presup/genero/informe/informebn2017/informe.htm [Accessed 10 Apr. 2018].

Kaika, M. (2012). The economic crisis seen from the everyday. *City: Analysis of Urban Trends, Culture, Theory, Policy, Action*, 16(4), 422–430.

Karamessini, M. (2014). Structural crisis and adjustment in Greece. Social regression and the challenge to gender equality. In: M. Karamessini and J. Rubery (Eds.), *Women and austerity. The economic crisis and the future of gender equality*. London and New York: Routledge, 165–184.

Laghaei, J., Faghri, A. and Li, M. (2015). Impacts of home shopping on vehicle operations and greenhouse gas emissions. Multi-year regional study. *International Journal of Sustainable Development and World Ecology*, 23(5), 381–391.

Linebaugh, P. (2007). *The Magna Carta manifesto. Liberties and commons for all*. Berkeley: University of California Press.

Lyberaki, A. and Tinios, P. (undated). *The Greek crisis and the return of the male breadwinner model? Some thoughts*. [Online]. Available at: www.ingenere.it/sites/default/files/articoli/Lyberaki-Tinios.pdf [Accessed 3 Mar. 2018].

Manganara, I. (2014). The effects of the economic crisis on women in Europe. *International Alliance of Women*, [Online] 4 December 2014. Available at: http://womenalliance.org/the-effects-of-the-economic-crisis-on-women-in-europe [Accessed 3 Mar. 2018].

McDowell, L. (1999). *Gender, identity and place. Understanding feminist geographies*. Cambridge, UK: Polity.

McLaren, D. and Agyeman, J. (2015). *Sharing cities. A case for truly smart and sustainable cities*. Cambridge, MA: MIT Press.

Morrow, K. (2017). Integrating gender issues into the global climate change regime. In: S. Buckingham and V. Le Masson (Eds.), *Understanding climate change through gender relations*. London: Routledge, 31–44.

Ostrom, E. (1990). *Governing the commons. The evolution of institutions for collective action*. Cambridge, UK: Cambridge University Press.

Peck, J. (2012). Austerity urbanism. American cities under extreme economy. *City*, 16(6), 626–655.

Rubery, J. and Tarling, R. (1982). Women in the recession. In: D. Currie and M. Sawyer (Eds.), *Socialist economic review*. Manchester: Merlin Press.

Sánchez de Madariaga, I. (2013). Mobility of care. Introducing new concepts in urban transport. In: I. Sánchez de Madariaga and M. Roberts (Eds.), *Fair shared cities. The impact of gender planning in Europe*. Farnham: Ashgate, 33–48.

Seager, J. (2009). Death by degrees. Taking a feminist hard look at the 2° climate policy. *Women, Gender and Research*, 3–4, 11–21.

Seguino, S. (2009). The global crisis, its gender implications and policy responses. *Gender Perspectives on the Financial Crisis Panel at the Fifty Third Session of the Commission on the Status of Women, United Nations*. [Online]. Available at: www.uvm.edu/~sseguino/pdf/global_crisis.pdf [Accessed 5 Mar. 2018].

Senatsverwaltung für Stadtentwicklung (Ed.) (2011). *Gender mainstreaming in urban development*. Berlin: Senatsverwaltung für Stadtentwicklung.

Svarna, F. (2014). *Financial crisis and domestic violence – the case of Greece*. The Wip Monterey Institute for International Studies. [Online]. Available at: http://thewip.net/2014/05/29/financial-crisis-and-domestic-violence-the-case-of-greece/ [Accessed 24 May 2014].

Tapsell, S.M. and Tunstall, S.M. (2001). *The health and social effects of the June 2000 flooding in the northeast region*. Report to the Environment Agency. Enfield: Flood Hazard Research Centre, Middlesex University.

Tapsell, S.M., Tunstall, S.M. and Wilson, T. (2003). *Banbury and Kidlington. Four years after the flood: An examination of the long-term health effects of flooding*. Report to the Environment Agency, Thames Region. Enfield: Flood Hazard Research Centre, Middlesex University.

Tummers, L. (2017). Co-housing. A double shift in roles? In: S. Buckingham and V. Le Masson (Eds.), *Understanding climate change through gender relations*. London: Routledge, 239–256.

UK Ministry of Housing, Communities and Local Government (2014). *Ensuring the vitality of town centres*. [Online]. Available at: www.gov.uk/guidance/ensuring-the-vitality-of-town-centres [Accessed 12 July 2018].

UK Parliament Home Affairs Committee (2013). *Written evidence submitted by Women for Refugee Women, the London Refugee Women's Forum and Women Asylum Seekers Together*. [Online]. Available at: https://publications.parliament.uk/pa/cm201314/cmselect/cmhaff/71/71we-4.htm [Accessed 3 Mar. 2018].

UNFCCC – United Nations Framework Convention on Climate Change (2015). *Paris Agreement*. [Online]. Available at: http://unfccc.int/files/essential_background/convention/application/pdf/english_paris_agreement.pdf [Accessed 3 Mar. 2018].

UNHCR – United Nations High Commissioner for Refugees (2000). *Reception standards for asylum seekers in the European Union*. Geneva, July 2000. [Online]. Available at: www.refworld.org/docid/3ae6b3440.html [Accessed 3 Mar. 2018].

UNICRI – United Nations Interregional Crime and Justice Research Institute, Emergent Crimes Unit Team (2014). *The impacts of the crisis on gender equality and women's wellbeing in EU Mediterranean countries*. [Online]. Available at: www.unicri.it/news/files/VAW_draft_last_lowq.pdf [Accessed 3 Mar. 2018].

Urbanistas (undated). *Urbanistas London*. [Online]. Available at: https://urbanistasuk.wordpress.com/urbanistas-london-3/ [Accessed 20 July 2017].

Vaiou, D. (2014a). Is the crisis in Athens (also) gendered? Facets of access and (in)visibility in everyday public spaces. *City: Analysis of urban trends, culture, theory, policy, action*, 18(4–5), 533–537.

Vaiou, D. (2014b). *Tracing aspects of the Greek crisis in Athens. Putting women in the picture*. European Urban and Regional Studies.

Vaiou, D. and Kalantides, A. (2016). Practices of collective action and solidarity. Reconfigurations of the public space in crisis-ridden Athens, Greece. *Journal of Housing and the Built Environment*, 31(3), 457–470.

Vaiou, D. and Lafazani, O. (2015). Kypseli and its market. Conflict and coexistence in the neighbourhoods of the city centre. *Athens Social Atlas*. [Online] December 2015. Available at: www.athenssocialatlas.gr/en/article/kypseli/ [Accessed 9 Apr. 2018].

Walby, S. (2009). *Gender and the financial crisis*. Paper for UNESCO project on "Gender and the financial crisis". [Online]. Available at: www.lancaster.ac.uk/fass/doc_library/sociology/Gender_and_financial_crisis_Sylvia_Walby.pdf [Accessed 3 Mar. 2018].

WEN – Women's Environmental Network (undated). *Healthier, cheaper, more sustainable – food growing brings people together*. Local Food Campaigns. [Online]. Available at: www.wen.org.uk/local-food/ [Accessed 31 Oct. 2018].

WHO – World Health Organisation (2010). *Gender, climate change and health*. Draft discussion paper. [Online]. Available at: www.who.int/globalchange/publications/reports/final_who_gender.pdf [Accessed 3 Mar. 2018].

Wrigley, N. and Lambiri, D. (2014). *High street performance and evolution. A brief guide to the evidence*. Southampton: University of Southampton/ESRC.

9 Conclusions and perspectives

Barbara Zibell, Doris Damyanovic and Ulrike Sturm

At the end of this book, we want to return to the question raised at the beginning: How does recognising the relevance of gender impact on the organisation and structure of land-use planning and appropriation in different European cultures? This question cannot, of course, be considered without referring to the global (and local) requirements of sustainable development.

In search of answers we first looked at definitions of and debates around the concepts and terms of gender, space and development, we retraced the genesis of urban respectively spatial planning in Europe since the nineteenth century, and we dealt with contradictions and challenges in transferring conceptions and political requirements in practice. Second, we presented case studies at a strategic/city-wide level and on local/neighbourhood level to determine the degree to which gender concepts have been implemented in planning practices, and discussed the challenges of conducting evaluations of planning and development projects from a gender+ perspective. Third, we reflected on supranational policies and their impact on national and local levels, discussing problems of global reach that create particular challenges for spatial development: climate change, migration and austerity. What is the current position? Which answers could we find to the questions initially raised? How satisfying are these answers? What has been achieved? What needs improvement? What has to be done next?

Conclusions – findings of gender in spatial development

Without any claim to completeness, we have been able to provide some answers; we have promoted the cross-national exchange of discourses on gender and spatial development and we have stimulated transcultural discourse. Looking at existing knowledge and at experiences with the transfer into practice of gender in spatial development – especially in Germany and some parts of Austria, where theoretical approaches and corresponding projects have existed for decades and are also quite well documented – we find that the mainstream (of development and planning practice as well as research) is still not permeated by gender sensitivity. This is the case even though theoretical concepts and discourses – for example as initiated by political guidelines – have influenced individual practical projects here and there and have found receptive contexts. These experiences have also been documented and evaluated. In general, it can be concluded that, even though gender mainstreaming (GM) was introduced in all EU countries in the Amsterdam Treaty of 1997 (adopted by Switzerland in 1999), the degree and intensity of its transfer, integration or implementation differ greatly between the eight EU countries represented here – or even within these countries. The cities and regions examined in the case studies (Chapters 4–7) are not representative of the countries as a whole, but are mostly exceptions in the country concerned.

The findings of the book can be summarised with regard to knowledge "transfer", knowledge "integration" and knowledge "implementation". By "transfer" we understand the mere practice, possibly applied only once, which in the best case makes gender issues visible in a lighthouse or pilot project, without lasting, structure-changing effects. The term "integration" is used to describe a greater sensitivity – including political sensitivity – which may be reflected, at least for a certain period of time, in local or supra-local/regional administrations and which influence the projects, concepts and planning measures of a city or region beyond the individual case. Finally, "implementation" means that the gender perspective is sustainably anchored in planning and development projects, processes and structures and is, as a matter of course, regularly adopted (and therefore no longer an explicit requirement). However, we are still miles away from this latter situation in all the cases considered and in all the knowledge of the countries represented in this book. The prerequisite for "sustainable" implementation would be political will that does not disappear in the next legislature, but has inscribed itself into the consciousness and self-conception of an entire generation. It would thus have a culturally transforming effect on the future of spatial development and society, that is to say it would be capable of bringing about genuine cultural change.

In this context Vienna, both the capital and a province of Austria, has clearly proven to be the model case in Europe, as further confirmed by long-standing exchanges among experts, some of whom were members of the IWG responsible for this book. With its social-democratic tradition, Vienna has been able to maintain its exemplary character in terms of urban development through social housing for almost a hundred years (Förster and Menking, 2016; see Chapter 5), with the consequence that a certain degree of self-evidence has been achieved in dealing with the gender perspective in planning. But even here, however, convincing personalities in more or less powerful positions are needed to repeatedly make the relevant demands. At least there are binding rules on the participation of women in decision-making bodies, where they are still in the minority, and on the inclusion of gender criteria in planning and measures in terms of content and processes, not only in housing construction and social infrastructure but also in mobility, public spaces and strategic spatial development. However, as a result mainstreaming gender and an explicit women's perspective in urban development were to a certain extent eliminated: the expert body of the City of Vienna, which had been set up specifically to represent the women's and everyday life perspective, was dissolved before firm integration and sustainable implementation could be achieved in all institutional structures. However, these gender experts are now employed in different planning units and still continue to work on implementation in urban planning (see Chapters 4, 5 and 7).

In other cities and regions of Europe we are still a long way from taking gender questions in spatial development and planning as a matter of course. In Berlin, Freiburg, Munich and in the Ruhr area, for example, there are approaches and some more or less established practices but they are far from being comprehensive or sustainably implemented. These approaches and practices are still rather delicate and fragile and have to be politically desired and enforced with conviction in order to thrive. The comparative studies from London, Zurich, Helsinki and other cities and regions (see Chapters 4–7) have clearly shown this for other spatial contexts as well. The studies have also highlighted differences in political guidelines and, above all, in the mental and cultural self-conception of the different European countries and their subregions. This became evident in the reception of the different chapters by the members of this international working group. Thus, to pick an example concerning planning regulations, the specification of a maximum building height or number of storeys has been overturned in some major cities in the UK, whereas it is conceivable in all German-speaking countries and is indeed regarded as an obvious instrument for the (top-down) control of the intensity and scope of spatial developments in Germany, Austria and Switzerland. Beyond all differences, however, a central common ground became apparent. The gender perspective, which ultimately aims to establish new power relations with respect to gender, divides opinions in all countries. Even if the same holds good for all countries, this "equality" comes in very different forms.

To summarise the findings with respect to the three levels of "transfer", "integration" and "implementation" we may conclude that experience on the first level of "transfer" is found to some extent in all the countries represented here. In Finland, the pioneer country for women's emancipation in Europe, the gender perspective in spatial planning and development does not seem to occur explicitly, thus implementation in the sense of a consciously adopted and anchored perspective has not (yet) taken place. In Switzerland, due to the political system of semi-direct democracy, there is no great tradition of explicitly anchored guiding principles (top-down), but there is a high level of quality awareness in general, which meets the concerns of gender-oriented spatial development – not only here and there, but also systematically in housing (see Chapter 5) and individual urban projects (see Cooperative Housing in Zurich, ibid.). The strongest tradition of an explicitly adopted perspective seems to exist in Germany and Austria, and more recently in Spain. In the Netherlands and Great Britain the gender perspective used to be more thoroughly implemented than it is today. However, there is no country in which developments have led to a situation that goes far beyond the stage of "transfer". In some cases, the stage of "integration" may be satisfied; however, there is as yet no recognition of the gender perspective as a continuous planning principle, which means that the level of "implementation" does not exist anywhere. And we can observe that the explicit mention of or demand for a gender perspective (by GM or other political guidelines) has in some cases led to a "wear and tear" effect: not only decision-makers have grown tired of the "gender debate", gender-related research and explicitly related practice are publicly "bashed".

Another finding of the book relates to the characterisation of similarities and differences of gender approaches in the (nine) different countries on a matrix using binary opposites (see Figure 9.1). As elaborated by Horelli (2017; see Figure 1.2 within this book), there are differences between cultures and practices on the two axes of formal/informal and top-down/bottom-up. During

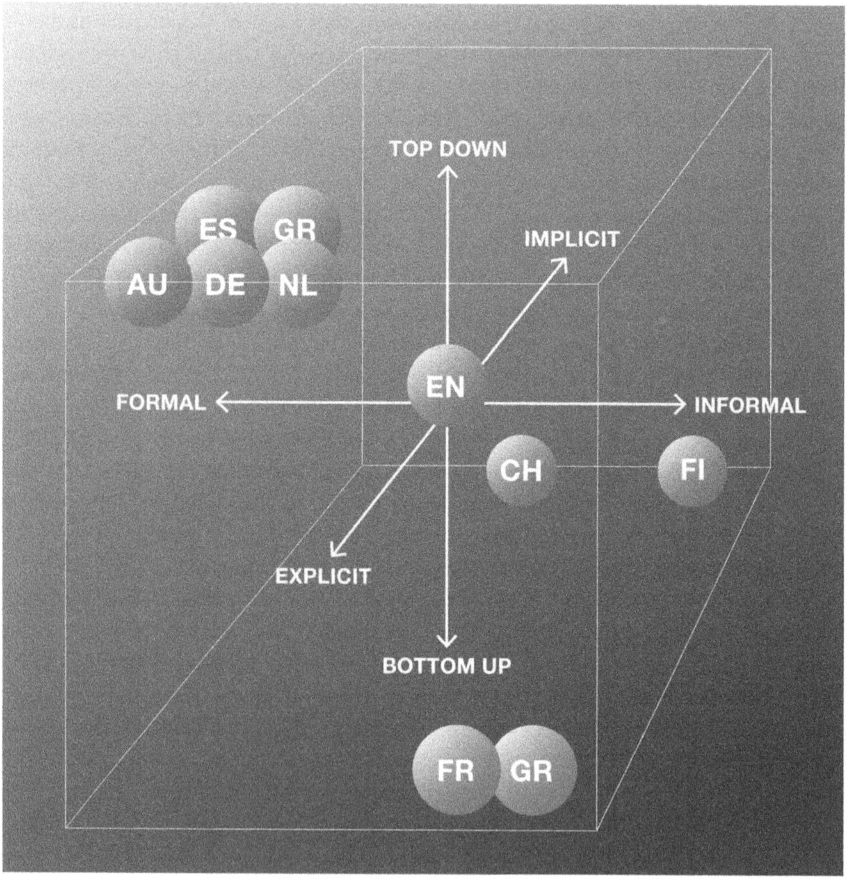

Figure 9.1 Characteristics of cases and cultures of gender implementation in the nine countries dealt with in the book.

Source: © Sturm, Zibell and Damyanovic.

the work on this book, we additionally found – not least by including Switzerland as a non-EU member – that there is another important axis: namely between implicit and explicit approaches. In some contexts, gender concerns are dealt with implicitly within quality-driven planning for everyday needs. But this is not sufficient if we aim to make gender sensibility a guiding principle of spatial development and planning. Explicit expression, insight and debate of strategic gender interests and practical gender needs have to precede any potential changing of planning cultures by transformed gender relations.

Greece has turned out to be a special case – and this obviously not only because of its special situation in view of the current austerity policy, but also as a result of its own mentality in earlier times (see Albers, 1997). Here there is both a strong tradition of formal building and planning legislation and a strong tradition of "nonetheless" cultures and self-help, which currently manifest themselves as survival strategies in various informal bottom-up activities in which women play a leading role. The country is therefore located in two positions in Figure 9.1. In general, it can be stated that formal, top-down and explicit approaches often go hand in hand with one another, as do informal, bottom-up and implicit approaches. But this is not necessarily so. The three-dimensional matrix turned out to be a useful instrument to characterise cases and cultures of gender – on the three levels of transfer, integration or (sustainable) implementation.

Perspectives – vision and requirements

What are the visions we are striving for? What kind of society do we envisage? And which spatial – physical as well as social – conditions are needed?

The notion of "good life" is currently much debated in public global and European discourse, as well as being documented in brochures and other media (see Chapter 7). Moreover, the idea and the right to a "good life" comprises a widespread scientific debate from philosophical, sociological and economic perspectives (e.g. Max-Neef, Elizalde and Hopenhayn, 1991; Nussbaum and Sen, 1993; Mies, 1995; Acosta, 2015), one that draws on ancient traditions. This debate has arrived in the political global arena, for instance in the Sustainable Development Goals (SDGs) and New Urban Agenda (see Chapter 7), and also on the national level, for example in Germany (Die Bundesregierung, 2016). A heightened awareness of the need for a different society with a higher degree of spatial and social justice (Fainstein, 2010) can thus be found.

If we – as a group of feminist-minded authors and editors – were to formulate a vision, we would imagine a society of people with equal rights, in which everybody would be able to take his or her place in society disregarding the gender they are. In this society each and every one of us could contribute to the sustainable use of space (and society) and its further development for the future.

What would such a space, such a city or region look or be like? And how could such a space be developed from the existing situation?

- Appearance – the "city shaped" (a notion used by Kostof, 1991): a town or village and any hybrids of them would be humane, which means: built and designed with respect to human dimensions, as living spaces for daily life and daily needs, i.e. differentiated in line with the various demands and needs in a given space. Accordingly, people would be integrated into planning projects and their implementation into practice.
- Composition – the "city assembled" (a notion also used by Kostof, 1992): a town or village and any hybrids of them would be well-adapted to the weakest member of society; slowness and short distances would be the focus of considerations; the planning for the appropriate structures of a humane city would be guided by the diverse needs of all genders.

And finally:

- Planning and development – the emergence of a city, town or village and any hybrids of them in democratic societies would be the concern of a multitude of people and a question of back-and-forth movements and mutual influences from bottom to top and from top to bottom, together forward, forward to far!

The challenges of the linkage of theory and practice

In its globally acclaimed report on the (necessary) "Great Transformation" (WBGU, 2011), the German Advisory Council on Global Change suggested, with reference to Polanyi (1944), a "new social contract" in order to achieve the necessary adaptation to climate change and to develop towards a resource-saving society. A "new social contract" is indeed necessary – but not without changed gender relations. Unfortunately, the report on the "Great Transformation" completely lacks a gender dimension. In our opinion, a "new social contract" cannot succeed without changed relations of power and control, and this also means without changed gender relations. The discourses on these topics are to be found in networks for caring economies – which have not yet arrived in the mainstream.

Ultimately, it is not only a question of the visibility and acceptance of bottom-up initiatives or of giving them a political voice and social weight – even if this is an important prerequisite for bringing about social transformation in the sense of real sustainable development. It is also a question of changing economic structures and how our livelihoods are created or restored. This also means to re-value unpaid work and work that can ensure the survival of all people in a sustainable, resource-saving and socially integrative and inclusive way.

This is not the place to draft out such a vision, there are already plenty of ideas – also in history and not least from women (see for example utopian feminists, Chapter 2 within this book). In fact, there is a lack of implementation, a lack of conviction that things guided by far-reaching visions can go well. And there is a lack of political will, which alone can lead to ideas and effective measures for the changes necessary to enable us to cope with climate change and migration and to avoid austerity politics at the expense of the weakest. Social innovation for the benefit of the poorest in world society *is* possible.

Formal, top-down and explicit approaches such as gender mainstreaming are a necessary ingredient in the way to change power relations, in addition to informal and bottom-up activities that may help to (re)introduce needs overlooked by official proceedings. Sustainable Development Goals or Gender Mainstreaming guidelines run the risk of not penetrating lower levels without more binding commitment in politics, law and planning. For this, the individual contexts of different cultures have to be taken into account, binding guidelines need to be developed in a context-specific and level-specific manner and introduced for each level of political action. Furthermore, responsibility has to be delegated to the local level. Measures like providing local communities with self-managed budgeting, allocating half of the resources and power to women at each level, have to be implemented as a matter of course.

The editors conclude that for knowledge integration and implementation, gender concerns principally have to be on the agenda of research and political programmes and should be part of training and education, especially in the disciplines relevant to spatial development like planning, architecture and urbanism. The authors are cautiously optimistic that the fourth generation of feminism will bring about more gender awareness and help to equip the next generation of professionals in these fields with the appropriate skills and tools for a truly gendered society, incorporating gendered spatial development and planning.

Gender research as a contribution to innovation and transformation

The authors use the summarised results of this book to highlight three important aspects for the successful integration and implementation of gender perspectives in planning research, theory and practice in spatial development and planning (Damyanovic, 2016).

Gender as a "critical perspective" in planning practice and in teaching and research supports a transformative approach concerning power relations and their influence on access to spaces, resources and decision-making processes by critically analysing, questioning and designing. Gender is still one of the key categories for discrimination and one which has proven to be very resistant in space and over time. Therefore "it must not be enmeshed into more widely diverse policy covering other factors of social disadvantage" (Sánchez de Madariaga, 2013: 330).

Gender as an "analytic tool" contributes to a deeper and more comprehensive understanding of the production of space and planning processes. Moreover, it supports the further development of theories, methodology and procedures in spatial development and planning. Gendered approaches in theory and practice of planning and development in principle critically review findings in research as well as in any scientific context (see Chapters 1, 2 and 6).

Gender as a "value" is a key structuring factor of society. The authors believe that it will contribute to more sustainable development outcomes and critical planning cultures. With such a value-based attitude, supporting a plurality of genders, identities and intersectionalities in thinking and acting would be a matter of course. Working towards gender equity and equality by integrating the symbolic dimension in research and planning practice and education is a prerequisite for changing planning models and the built environment (Damyanovic, 2016). For this reason, it is absolutely necessary to implement gender as an additional value into spatial development and planning, by applying gender-specific knowledge and criteria in every spatial context and measure.

References

Acosta, A. (2015). *Buen vivir. Vom Recht auf ein gutes Leben (Good living. About the right to good life)*. München: oekom verlag.

Albers, G. (1997). *Zur Entwicklung der Stadtplanung in Europa. Begegnungen, Einflüsse, Verflechtungen (The development of urban planning in Europe. Encounters, influences, interdependencies)*. Bauwelt Fundamente 117. Braunschweig/Wiesbaden: Vieweg.

Damyanovic, D. (2016). *A gender-sensitive approach in landscape planning. Theoretical and methodological prerequisites and findings of basic and applied research to develop further a gender-sensitive approach in landscape planning*. Professorial Thesis (Habilitationsschrift). University of Natural Resources and Life Sciences, Vienna.

Die Bundesregierung (German Federal Government) (Ed.) (2016). *Bericht der Bundesregierung zur Lebensqualität in Deutschland (Report of the Federal Government about life quality in Germany)*. Berlin. [Online]. Available at: www.gut-leben-in-deutschland.de/static/LB/index.html [Accessed 29 Oct. 2018].

Fainstein, S.S. (2010). *The just city*. New York: Cornell University Press.

Förster, W. and Menking, W. (Eds.) (2016). *Das Wiener Modell. Wohnbau für die Stadt des 21. Jahrhunderts (The Vienna Model. Housing for the twenty-first-century city)*. Berlin: Jovis.

Horelli, L. (2017). Engendering urban planning in different contexts – successes, constraints and consequences. *European Planning Studies*, 25(10), 1779–1796.

Kostof, S. (1992). *The city assembled. The elements of urban form through history*. Boston, Toronto, London: Little, Brown and Company.

Kostof, S. (1991). *The city shaped. Urban patterns and meanings through history*. London: Thames and Hudson.

Max-Neef, M.A., Elizalde, A. and Hopenhayn, M. (1991). *Human scale development. Conception, application and further reflections*. New York: The Apex Press.

Mies, M. (1995). Befreiung vom Konsum. Subsistenz: Freiheit oder Liberalisierung (Liberation from consumption. Subsistence: freedom or liberalisation). In: M. Mies and V. Shiva (Eds.), *Ökofeminismus (Ecofeminism)*. Zürich: Rotpunkt.

Nussbaum, M.C. and Sen, A. (Eds.) (1993). *The quality of life*. Oxford: Clarendon Press.
Polanyi, K. (1944). *The great transformation*. New York: Farrar & Rinehart.
Sánchez de Madariaga, I. (2013). Looking forward, moving beyond trade-offs. In: I. Sánchez de Madariaga and M. Roberts (Eds.), *Fair shared cities. The impact of gender planning in Europe*. Farnham: Ashgate, 325–333.
WBGU – German Advisory Council on Global Change (Eds.) (2011). *World in transition – a social contract for sustainability*. [Online] Berlin. Available at: www.wbgu.de/en/flagship-reports/fr-2011-a-social-contract// [Accessed 29 Oct. 2018].

Index

www.ingramcontent.com/pod-product-compliance
Ingram Content Group UK Ltd.
Pitfield, Milton Keynes, MK11 3LW, UK
UKHW010958180926
475701UK00004B/428
* 9 7 8 0 3 6 7 7 3 0 7 9 6 *